CARE-W

Computer Aided Rehabilitation for Water Networks

CARE-W

Computer Aided Rehabilitation for Water Networks

Edited by

Sveinung Saegrov

IWA Publishing

LONDON • SEATTLE

Published by IWA Publishing, Alliance House, 12 Caxton Street, London SW1H 0QS, UK

Telephone: +44 (0) 20 7654 5500; Fax: +44 (0) 20 7654 5555; Email: publications@iwap.co.uk
Web: www.iwapublishing.com

First published 2005

British Library Cataloguing in Publication Data
A CIP catalogue record for this book is available from the British Library

Library of Congress Cataloging- in-Publication Data
A catalog record for this book is available from the Library of Congress

ISBN: 1843390914
ISBN: 9781843390916

Contents

Foreword

CARE-W was a project under the fifth Framework Programme of the European Commission from 2001 to 2004. The project had a budget of 3 million Euro. It included eleven partners from eight countries in Europe. Additionally twelve cities acted as end users. The project has produced a decision support system for water companies to use when planning upgrading of water networks. This report comprises a concentrated description of the methods used, the tools and software produced and the results from using it. The testing has clearly demonstrated that this system has a potential to support a proactive rehabilitation strategy.

The water network planner will hopefully see the CARE-W logo and name in many years from now. It is our ambition that the CARE-W system will be a leading instrument for the water industry in upgrading their water networks and for general management of the networks. We intend to establish an organisation to link the international companies and institutes working with CARE-W. At least from the beginning the end users will meet the current CARE-W partners as the CARE-W service providers.

We would like to thank the partners and the end users who have participated in the CARE-W project. Without the end users, it would not have been possible to complete this project in practice. Without the scientific competence of the partners, CARE-W could never have been realized. CARE-W has not been a straightforward exercise. It relies heavily on many years of research and includes many doctoral degrees. The CARE-W project has built on existing scientific work and developed new elements, to provide a complete system for water network rehabilitation.

We would also like to thank the European Commission and the administration of the Fifth Framework Programme that made CARE-W possible by its financial support. Additionally, we thank the national agencies who provided the remaining resources for the fulfilment of CARE-W.

Sveinung Sægrov
CARE-W co-ordinator
Trondheim July 2005

Summary

GENERAL

CARE-W is a joint European initiative to develop a framework for water network rehabilitation. The project has been supported by the European Commission under the Fifth Framework Programme for Research and Development and had a budget of 3 million Euro.

The aim of CARE-W was to support European water companies in their decisions on upgrading their water supply. The system has been developed for and is tested by cities representing all parts of Europe. The project group is now ready to use it for practical support to end users.

THE CARE-W PARTNERSHIP

Figure 1 below shows the partnership of CARE-W. 11 research partners and 13 end users were involved, representing 8 countries from all parts of Europe. CARE-W was coordinated by SINTEF, Norway. The idea of a comprehensive European planning system for water network rehabilitation was first presented in 1999 within the COST C3 action on urban water infrastructure. The proposed project was approved by the European Commission and started in 2001, as a three year project. The project was finished 31. January 2004. The CARE-W project partners were:

Table 1: CARE-W project partners

Partner	Contact person
Foundation for Industrial and Technical Research, SINTEF, Trondheim, Norway	Prof. Sveinung Sægrov Prof Wolfgang Schilling
Norwegian University of Science and Technology, NTNU, Trondheim, Norway	Dr. Jon Røstum
Brno University of Technology, Czech Republic	Dr. Ladislav Tuhovcak
Cemagref, Bordeaux, France	Dr. Patrick Eisenbeis
Dresden University of Technology, Dresden, Germany	Prof Raimund Herz
INSAVALOR, Lyon, France	Prof. Pascal LeGauffre
National Civil Engineering Laboratory, LNEC, Lisboa, Portugal	Dr. Helena Alegre
Water Research Centre, WRc, Swindon, United Kingdom	Dr. Paul Conroy
University of Bologna, Bologna, Italy	Prof Vittorio DiFederico
University of Ferrara, Ferrara, Italy	Prof Sante Mazzacane
AGAC, Reggio Emilia, Italy	Eng. Marcello Schiatti

The cities of Bristol/UK, Brno/CZ, Codigoro/I, Dresden/D, Ferrara/I, Lausanne/CH, Lyon/F, Oeiras Amadora/P (close to Lisbon), Oslo/N, Reggio Emilia/I, Roubaix Tourcoing/F, Stuttgart/D and Trondheim/D have acted as CARE-W end users, and have tested the system.

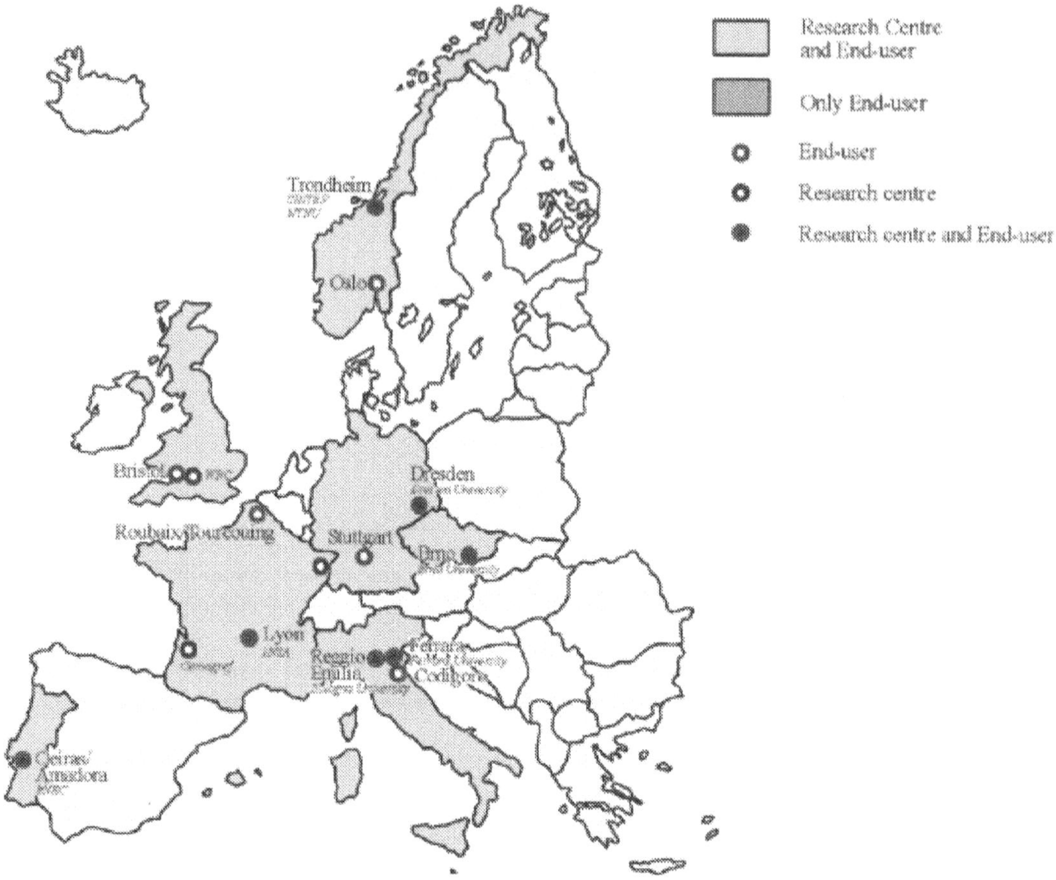

Figure 1 Partnership of CARE-W

The CARE-W system

CARE-W consists of software dealing with fundamental instruments for estimating the current and future condition of water networks, including tools to assess performance indicators (PI), to predict pipe failures (FAIL) and to calculate water supply reliability (REL). Based on the results of these tools, long-term investment needs are estimated (LTP tool) as well as annual rehabilitation projects selected and ranked (ARP tool). The tools are operated jointly within the "CARE-W Manager", which also contains facilities for using pipe network databases, geographical information systems (GIS) and input/output routines. The results from using the procedures are presented by reports, in tables and graphically.

The end user will operate CARE-W through a GIS-based interface. The system can be used for the analysis of an entire water network, a district or a cluster of pipes, defined by certain criteria. The user may run all the tools or a selection, depending on the problem to be solved. Generic statistics on the general performance of the water network may be obtained from the PI tool. The FAIL tool allows the probability of pipe breaks to be predicted and the REL tool combines this information with the hydraulic importance of the pipes, thus allowing the prediction of the service reliability. This information is fed into the ARP (Annual Rehabilitation Planning) tool that uses a multi-criteria selection procedure to rank different rehabilitation projects. Information on future investment needs relevant for master plans is generated with the LTP (Long Term Planning) tool.

The system is illustrated in the figure below. The main outcome of CARE-W when running the complete system is a set of selected and ranked rehabilitation projects and a long-term investment plan.

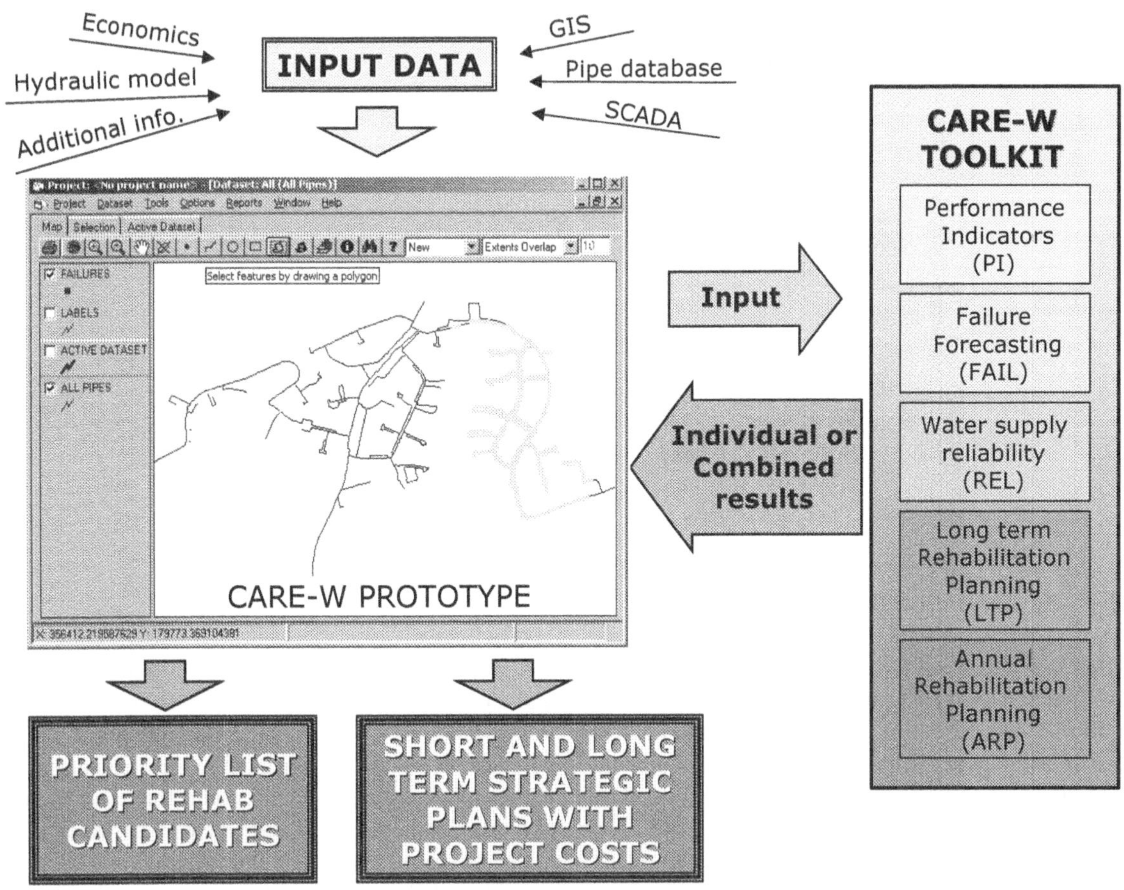

Figure 2 Overview of CARE-W system

The **CARE-W PI** (Performance Indicators) enables the network owner to measure the performance of the network with a range of key indicators. The CARE-W PI tool is based upon the IWA PI-system for drinking water. It consists of 49 Performance Indicators (PI), 154 Utility Indicators (UI) and 29 External Indicators (EI). The PI are relative numbers expressing resources spent (e.g. rehabilitation cost/km), condition (number of breaks/km) or performance (number of complaints per 1000 connections per year). The UI are data that are directly related to the activity of the utility and is under its direct control, for example: length of network. It is used either for the assessment of PI or for the CARE-W decision making process. The EI are data that cannot be directly influenced by the utility, but that is critical for establishing the rehab diagnosis or for support to the CARE-W decision-making process. Examples of EI data are rain, temperature, ground slope, type of soil.

The system is meant to be an appropriate tool for studies to fulfil objectives and analysis of effects of previous investments. PI also provides a source for benchmarking studies. Benchmarking studies were carried out in the context of the CARE-W project including 30 cities, and have served as a basis for establishing recommended values for several PI. Benchmarking between city districts can give useful information for rehabilitation strategies. In general, the CARE-W PI and benchmarking system can be very valuable for a water master plan and rehabilitation plans.

PI can be presented in different ways, e.g. pictorially using the GIS viewer, in tabular reports or by using in-built graphic features of the PI tool such as time series plots. Figure 3 shows time series plots of two PI. This particular example demonstrates the relationship between action (mains rehabilitation) and result (mains failures).

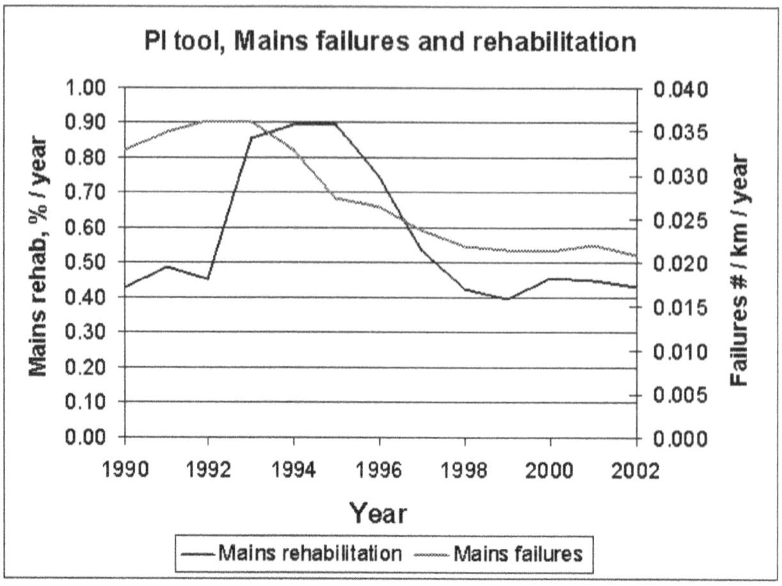

Figure 3 Example of CARE-W PI use: Time series plot

Software has been developed for the operation of the CARE-W PI tool. This software is fully integrated in the CARE-W Rehabilitation Manager but can also be used as a stand-alone tool.

The **annual rehabilitation planning system (CARE-W ARP)** presents a multi-criterion selection and ranking system that combines results from the CARE-W tools with additional information supplied by the user. This procedure enables analysis of the entire network, sectors or clusters of pipes. It provides recommendation of groups of pipes that should be considered for rehabilitation. Tools for prediction of future failures as well as water supply service reliability support CARE-W ARP. Projects can be selected based on failure frequencies, water quality problems as well as opportunity (street works for other reasons). The ARP analyses each project towards a number of criteria covering financial, technical as well as functional issues. Thereafter a general ranking system is set up and the projects are classified.

Avoiding failures is one criterion of rehabilitation, i.e. avoiding damage and loss of water supply following bursts and leaks. A set of statistical tools has been developed and tested to obtain a probabilistic forecast of such failures. They are included in the system as **CARE-W FAIL**. The underlying procedure also quantifies the effect of external conditions to the failure probability, such as soil, material, construction year and number of previous failures. It has been shown that the number of previous failures is the most sensitive parameter for failure forecasting. It has also been demonstrated that these tools can provide extra information if the data records are suffering from missing data or only cover a short registration period. The quality of the data records is, however, still critical for, both, an efficient use of CARE-W as well as management in general. Principles and methods for improving data quality were developed within CARE-W.

A combined hydraulic/reliability model, **CARE-W REL**, can analyze the loss of water supply due to bursts and leaks. A commercial hydraulic model, like EPANET, is combined with a routine forecasting the probability of failure for each pipe. The joint model considers the probability of leak or bursts as well as reduction of flow capacity and the consequences for water supply measured as flow and pressure to the customers. The model identifies "reliability bottlenecks" in the network. It has been used by cities to assess the reliability of water supply to sensitive customers and entire water supply districts. Figure 4 shows an example from a CARE-W REL analysis, where the bottlenecks are shown in black.

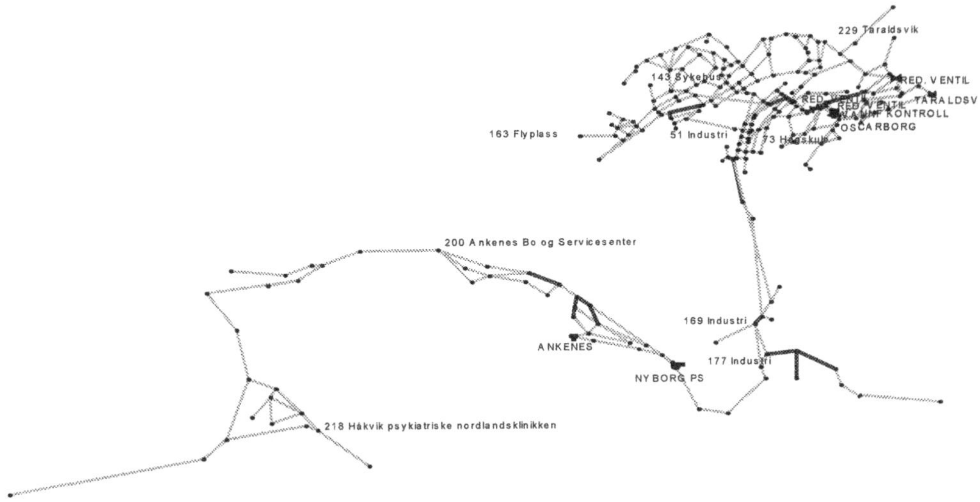

Figure 4 Example from the use of CARE-W REL. Finding "reliability bottlenecks" of water through combination of hydraulic capacity and failure probability

The **long-term planning** module, **CARE-W LTP** analyses the necessary investment level in the coming decades and how this is influenced by different rehabilitation strategies. The LTP module consists of three closely related tools, namely the Scenario Writer for developing consistent scenarios, the Rehab strategy manager (based on KANEW software) for simulation of long-term effects of specific rehabilitation options and finally the Rehab Strategy Evaluator that supports the selection of the best long-term rehabilitation strategy.

The module is based on statistics for asset survival and qualified assessment of survival for categories of pipes. The pipe material, the construction period and the diameter are the most important criteria for classification. The result is a general recommendation of network exchange for each pipe category. Some tests have indicated a rehabilitation need of 0.5-1% per year, but this number will highly depend on network history, climate, pipe depth and other technical issues.

All end users tested the CARE-W tools extensively. They have tested the functionality and analysed how it fits to their planning needs. The city of Oslo is among the test cases where results derived from applying the CARE-W tools have been compared to the priorities given by Oslo's traditional way of planning. Some similarities could be seen, but also some differences. In this case CARE-W sometimes confirmed but sometimes also challenges traditional thinking. The evaluation concluded that Performance Indicators, the Annual Rehabilitation Planning and the Long-Term Planning tools will be important elements in future rehabilitation planning and asset management in Oslo.

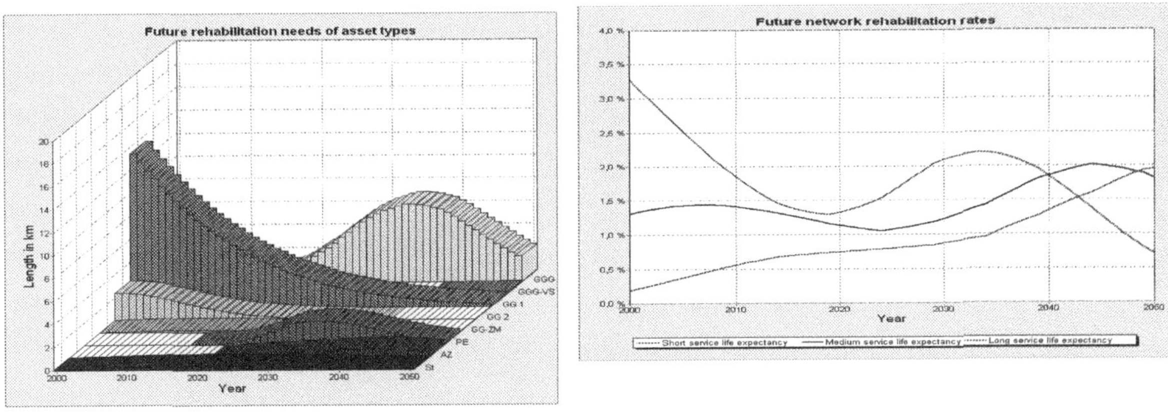

Figure 5 Rehab needs pr pipe materials and accumulated rehabilitation rates for "optimistic" and "pessimistic" service life perspective

THE MARKET FOR CARE-W

The main objective of CARE-W is to enable the utility manager to rehabilitate the right pipe at the right time. The basic requirements are to avoid premature rehabilitation (i.e. rehabilitation of the wrong pipe), interruption of water supply (i.e. due to unexpected pipe break), and also to avoid poor water quality. This should be done in a cost-efficient manner. CARE-W is a tool to define the right priority of projects, to avoid unnecessary investments, and to avoid negative financial and social consequences of major bursts and leaks.

There are approximately 2 million km of water pipes owned by public or private utilities in Europe. A similar length of private house connections also exists. CARE-W does not (yet) targets these. The annual rehabilitation need can be estimated to be at least 0.5%, i.e. 10.000 km pr year. The cost of this is 5 billion Euros. If decision support based on CARE-W increases the efficiency of this work by 20% (choosing the right pipes, avoiding consequences of pipe failures), the annual saving will be 1 billion Euros for all European cities. If CARE-W is used worldwide, the potential savings will be 3-5 times higher.

The potential clients of CARE-W are owners of water supply systems, operating companies (utilities), financial institutions (e.g. European Development Bank) and national regulators. The first two are large by number. Each of them has their individual information requirements: The owners or shareholders of the assets are interested in maintaining the value of their infrastructure and to define rehabilitation standards in a contract with the utility that operates the system. The utility needs information on investment needs to maintain the assets with respect to the standard defined in the contract. The financial institutions and potential investors need information on the magnitude and economic viability of rehabilitation investments, and the regulator needs such information for making decisions about performance requirements and water prices.

Other types of "clients" are universities and consultants who will need courses and information on the CARE-W technology. They will create a greater awareness of CARE-W and promote indirectly within the wider scientific community.

To summarize, we believe the CARE-W approach (consultancy and software) will be useful for:

- Water system owners
- Water system operators
- Banks, other finance and investment organisations
- National regulators
- Universities (courses)
- Consulting companies (courses)

The CARE-W consortium has the ambition that the CARE-W system shall be a leading planning instrument for the water industry in upgrading their water networks and for general management of the networks. The sheer volume of existing water supply systems, its expected service time into the future, and the economic turnover for necessary network rehabilitation is attractive for developing commercial activities in this area. We wish that the system will be continuously updated and improved for at least another 20 years from now, and will take steps to make this happen

1

Performance Indicators for water network rehabilitation

Helena Alegre, Jaime Melo Baptista and Sergio Coelho, LNEC

1.1 FRAMEWORK

In the CARE-W water supply network rehabilitation procedure, a system of performance indicators (PI) has been developed specifically for pipe network rehabilitation (Figure 1.1). The system is inspired on the International Water Association (IWA) performance indicators system for water supply services (Alegre *et al.*, 2000) and is described in full detail in Appendix 2-4. The CARE-W rehab PI system includes performance indicators, additional performance measures, utility information and external information, established according to the following concepts:

- The ***performance indicators*** (PI) are ratios between values of identical or different nature, expressing the performance of the undertaking regarding a given point of view relevant in the rehabilitation framework.
- The ***additional performance*** measures (APM) are also ratios between values of identical or different nature, expressing the performance of the undertaking regarding a given point of view relevant in the rehabilitation framework, but more difficult to define and assess (e.g. requiring the use of simulation models or of other specific algorithms).
- The ***utility information*** (UI) is the set of data that is directly related to the activity of the utility (organization and its physical system) and is under its direct control. It is used either for the assessment of the selected PI (as PI input variables, as defined in the IWA-PI system) or for the CARE-W decision making process.
- The ***external information*** (EI) is the set of data that cannot be directly influenced by the utility (external to the organization and to its physical system) but that is critical for establishing the rehab diagnosis or for support to the CARE-W decision-making process (e.g. rainfall, temperature, ground slope, type of soil, etc.). This basically corresponds to the 'External information / Region profile' group of the IWA-PI system.

One of the essential features of this system is the fact that the user is encouraged to select and use only the PI that are relevant for the particular case.

In the current version of the CARE-W rehab PI system the user can choose from a set of 49 performance indicators, with the associated utility information (UI) used for their calculation or interpretation (a maximum of 154 variables). There are also up to 29 external information parameters that can be used for providing context background.

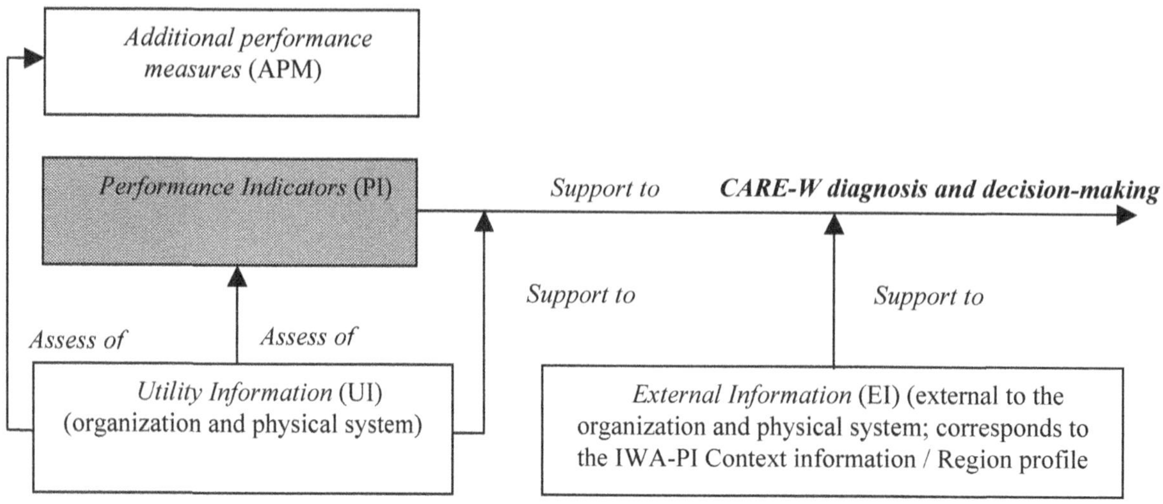

Figure 1.1. The CARE-W rehab PI system

1.2 COMPONENTS OF THE CARE-W PI SYSTEM

1.2.1 Performance indicators

The following PI groups have been identified as relevant for the CARE-W rehab system, and have been grouped in the same sections as the IWA-PI:

- operational indicators (rehabilitation, failures and repairs, water losses);
- quality of service indicators (service, customer complaints);
- financial indicators (annual costs, annual investments for network mains, tariffs);
- water resources indicators;
- physical indicators.

The most important group of rehab PI in this set is the Operational PI group, including not only the indicators that better express the alternative options to be adopted for rehabilitation, but also those related to failures and repairs and to water losses. The Quality of Service Indicators, particularly those related to service and customer complaints, and the Financial Indicators, including annual cost, annual investments and tariffs are equally rather important.

The rehab indicators selected comply with a good number of pre-set requirements, such as:

- to express the performance of the undertaking regarding a given point of view relevant in the rehabilitation framework;
- to represent all the relevant mains rehabilitation aspects of a water undertaking's performance;
- to be suitable for representing those aspects in a true and unbiased way;
- to be clearly defined, with a concise meaning and a unique interpretation for each indicator;
- to include only non-overlapping performance indicators;
- to be ratios between values of identical or different nature;
- to require only measuring equipment that is affordable, the need for sophisticated and expensive equipment being avoided;
- to be auditable, which is specially important when the performance indicators are to be used by regulatory bodies that may need to check the results reported;
- to be easy to understand, even by non-specialists e.g. consumers, wherever possible.

Table 1.1 presents the full listing of CARE-W rehab performance indicators, including their designations and units of expression. A clear processing rule for each PI and the definition of all the UI and EI are fully defined in Appendix 2-4.

1.2.2 Additional performance measures

As mentioned before, further to this set of PI it was decided to create a group of additional performance measures, which can be useful for the CARE-W project but are more difficult to be defined and assessed, do not guarantee unbiased and whose thresholds are not easy to establish. Some of these measures are under developed for other modules of the CARE-W Rehabilitation Manager by other partners of the project, according to their own needs and based on their specific tools, such as hydraulic modelling or statistical analysis. Some potential examples of additional performance measures are network hydraulic reliability, mains residual service life, service connections residual service life, balance of costs and benefits and internal rate of return.

Table 1.1. Full listing of the CARE-W rehab PI

DESIGNATION	DESIGNATION
OPERATIONAL INDICATORS	**QUALITY OF SERVICE INDICATORS**
Rehabilitation	**Service**
Mains rehabilitation (%/year)	Pressure of supply adequacy (%)
mains relining (%/year)	Water interruptions (%)
replaced or renewed mains (%/year)	Interruptions per connection (No./1000 connections)
replaced or renewed valves (%/year)	critical interruptions per connection (No./1000 connections)
Service connection rehabilitation (%/year)	Population experiencing restrictions to water service (%)
Failures and repairs	Days with restrictions to water service (%)
Mains failures (No./100 km/year)	Quality of supplied water (%)
pipe failures (No./100 km/year)	aesthetic test compliance (%)
point failures (No./100 km/year)	water taste test compliance (%)
valves failures (No./100 km/year)	water colour test compliance (%)
service connection insertion point failures (No./100 km/year)	microbiological test compliance (%)
Critical mains failures (No./100 km/year)	physical-chemical test compliance (%)
Service connection failures (No./1000 connect./year)	**Customer complaints**
Hydrant failures (No./1000 hydrants / year)	Service complaints per connection (No. complaints/ 1000 connection/ year)
Power failures (hours/ pumping station/year)	pressure complaints (%)
Active leakage control repairs (No./100 km /year)	continuity complaints (%)
Water losses	water quality complaints (%)
Water losses (m^3/connection/ year)	water taste complaints (%)
real losses (l/connection/ day)	water colour complaints (%)
Infrastructure leakage index (-)	interruptions complaints (%)
	critical interruptions complaints (%)
FINANCIAL INDICATORS[1]	**WATER RESOURCES INDICATORS**
Annual costs	Inefficiency of use of water resources (%)
Unit total costs (€/m^3)	Annual water resources availability ratio (%)
unit running costs (€/m^3)	**PHYSICAL INDICATORS**
Annual investment for network mains	Transmission and distribution storage capacity (days)
Unit investment for network mains (€/m^3)	Valve density (No./km)
annual investments for new and upgrading mains (%)	
annual investments for mains replacement (%)	
Tariffs	
Average water charges for direct consumption (€/m^3)	
Average water charges for exported water (€/m^3)	

1.2.3 Utility information

The UI variables are either needed to assess the selected PI (as PI input variables) or accessory in the CARE-W decision-making process. They are grouped in six sections (the same adopted by IWA):

- physical assets data - distribution network (transmission and distribution network, water storage, pumping stations);
- physical assets data - service connection;
- water volume data;
- operational data (service pressure, service continuity, water quality monitoring, inspection and maintenance, preventive maintenance, failures, rehabilitation);
- quality of service data (service, customer complaints);
- financial data (annual costs, annual investments for network mains, annual revenues, unit costs, tariffs).

The CARE-W PI system includes the following UI variables:

[1] In the IWA PI manual the monetary unit adopted is US$.

DESIGNATION	Unit	Code
PHYSICAL ASSETS DATA - DISTRIBUTION NETWORK		
Transmission and distribution network		
Mains length:		
Mains length	km	C6
Mains length in sensitive areas	km	C6a
Network extension	km/year	C6b
Mains diameters:		
mains with < 100/110 mm diameter	km	C15
mains with 100/110 < diameter ≤ 300/315 mm	km	C16
mains with diameter > 300/315 mm	km	C17
Mains materials:		
cast, spun & grey iron mains	km	C7
ductile iron mains	km	C8
steel mains	km	C9
asbestos cement mains	km	C10
polyethylene mains	km	C11
polyvinyl chlorine mains	km	C12
concrete mains	km	C13
other material mains	km	C14
Mains protection:		
mains with internal protection	km	C34
mains with external protection	km	C35
mains with cathodic protection	km	C36
Mains age:		
average mains age	years	C37
mains laid between 2006 and 2010	km	C18a
mains laid between 2001 and 2005	km	C18b
mains laid between 1996 and 2000	km	C18c
mains laid between 1991 and 1995	km	C18d
mains laid between 1986 and 1990	km	C18e
mains laid between 1981 and 1985	km	C18f
mains laid between 1976 and 1980	km	C18g
mains laid between 1971 and 1975	km	C19a
mains laid between 1966 and 1970	km	C19b
mains laid between 1961 and 1965	km	C19c
mains laid between 1956 and 1960	km	C19d
mains laid between 1951 and 1955	km	C19e
mains laid between 1946 and 1950	km	C20a
mains laid between 1941 and 1945	km	C20b
mains laid between 1936 and 1940	km	C20c
mains laid between 1931 and 1935	km	C20d
mains laid between 1926 and 1930	km	C20e
mains laid between 1921 and 1925	km	C21a
mains laid between 1916 and 1920	km	C21b
mains laid between 1911 and 1915	km	C21c
mains laid between 1906 and 1910	km	C21d
mains laid between 1901 and 1905	km	C21e
mains laid before 1900	km	C21f
Mains location:		
under flexible roadway	km	C38
under rigid roadway	km	C39
under sidewalk	km	C40
under green areas	km	C41
Mains installation depth	m	C42
Bedding type:		
mains with appropriate bedding	km	C43
mains with inadequate bedding	km	C43a
Backfilling soil type:		
mains with very aggressive backfilling soil	km	C44
mains with medium aggressive backfilling soil	km	C44a
mains with low aggressive backfilling soil	km	C44b
mains with not aggressive backfilling soil	km	C44c
Average closeness to trees	m	C45
Type of joints:		

DESIGNATION	Unit	Code
pipe rigid joints	km	C46
pipe flexible joints	km	C47
Main valves	No.	C29
Isolating valves	No.	C30
Hydrants	No.	C31
Water storage		
Transmission & distribution storage tanks capacity	m^3	C2
Pumping stations		
Pumping stations	No.	C4
PHYSICAL ASSETS DATA - SERVICE CONNECTIONS		
Number of service connections	No.	C32
Number of sensitive service connections	No.	C32a
Average service connection length to measurement point	m	C33
Service connection materials:		
Plastic service connections	%	C48a (IWA CI)
Steel service connections	%	C48b (IWA CI)
Lead service connections	%	C48c (IWA CI)
Other service connections	%	C48d (IWA CI)
WATER VOLUME DATA		
Annual abstraction capacity	m^3/year	A1
Imported water allowance	m^3/year	A2
Water abstracted	m^3/year	A4
Imported raw water	m^3/year	A5
Exported raw water	m^3/year	A6
Water produced	m^3/year	A7
Imported treated water	m^3/year	A8
Exported treated water	m^3/year	A9
Authorized consumption	m^3/year	A19
Water losses	m^3/year	A20
Real losses	m^3/year	A24
OPERATIONAL DATA		
Service pressure		
Average operating pressure	kPa	D31
Pressure variation	kPa	D31a (IWA CI)
Service continuity		
Time system is pressurised	h/year	D29
Delivery points with adequate pressure	No./year	D30
Water interruptions	No.h/year	D32
Service interruptions	No./year	D33
critical interruptions	No./year	D33a
Water use restrictions	No.h/year	D34
Days with restrictions to water service	days/year	D35
Water quality monitoring		
Water quality tests performed	No./year	D41
Aesthetic tests performed	No./year	D42
taste tests performed	No./year	D42a
colour tests performed	No./year	D42b
Microbiological tests performed	No./year	D43
Physical-chemical tests performed	No./year	D44
Compliance of aesthetic tests	No./year	D51
compliance of taste tests	No./year	D51a
compliance of colour tests	No./year	D51b
Compliance of microbiological tests	No./year	D52
Compliance of physical-chemical tests	No./year	D53
Inspection and maintenance		
Network inspection	km/year	D7
Leaks repaired due to active leakage control	No./year	D9
Preventive maintenance		
Mains rehabilitation	km/year	D18
mains relining	km/year	D19
replaced or renewed mains	km/year	D20
replaced or renewed valves	No./year	D21
Service connection rehabilitation	No./year	D22

DESIGNATION	Unit	Code
Failures		
Mains failures	No./year	D25
pipe failures	No./year	D25a
joint failures	No./year	D25b
valve failures	No./year	D25c
service connection insertion point failures	No./year	D25d
Mains failures caused by third parties	No./year	D25e
Critical mains failures	No./year	D25f
Service connection failures	No./year	D26
Hydrant failures	No./year	D27
Power failures	h/year	D28
Rehabilitation		
Mains rehabilitated	km/year	D36
mains relined	km/year	D36a
replaced or renewed mains	km/year	D36b
replaced or renewed valves	km/year	D36c
Mains rehabilitation average date	year	D37
mains relining average date	year	D37a
replaced or renewed mains average date	year	D37b
replaced or renewed valves average date	year	D37c
Service connection rehabilitated	No./year	D38
Service connection rehabilitation average date	year	D39
QUALITY OF SERVICE DATA		
Service:		
Population supplied	No. persons	F1
Population supplied with service pipes	No. persons	F2
Customer complaints:		
Service complaints	No./year	F11
pressure complaints	No./year	F12
continuity complaints	No./year	F13
water quality complaints	No./year	F14
water taste complaints	No./year	F14a
water colour complaints	No./year	F14b
complaints on interruptions	No./year	F15
complaints on critical interruptions	No./year	F15a
FINANCIAL DATA		
Annual costs:		
Annual costs	€/year	G1
Annual running costs	€/year	G2
Annual capital costs	€/year	G3
Annual depreciation costs	€/year	G19
Annual investments:		
Annual investment in tangible assets	€/year	G26
Annual investments for new assets	€/year	G27
Annual investments for assets replacement	€/year	G28
Annual revenues:		
Annual revenue	€/year	G21
Water sales revenue for direct consumption	€/year	G30
Water sales revenue for exported water	€/year	G31
Unit costs:		
Pipe rehabilitation cost	€/m	G51
Pipe repair cost	€/m	G52
Pipe maintenance cost	€/m	G53
Pipe inspection cost	€/m	G54
Reduced cost from pipe repair	€/year	G55
Reduced cost from leakage repair	€/year	G56
Tariffs:		
domestic water consumption tariff	€/m^3	G57 (IWA CI)
industrial water consumption tariff	€/m^3	G58
public water consumption tariff	€/m^3	G59

As shown in Figure 1.2, UI are grouped in six sections (the same adopted by IWA).

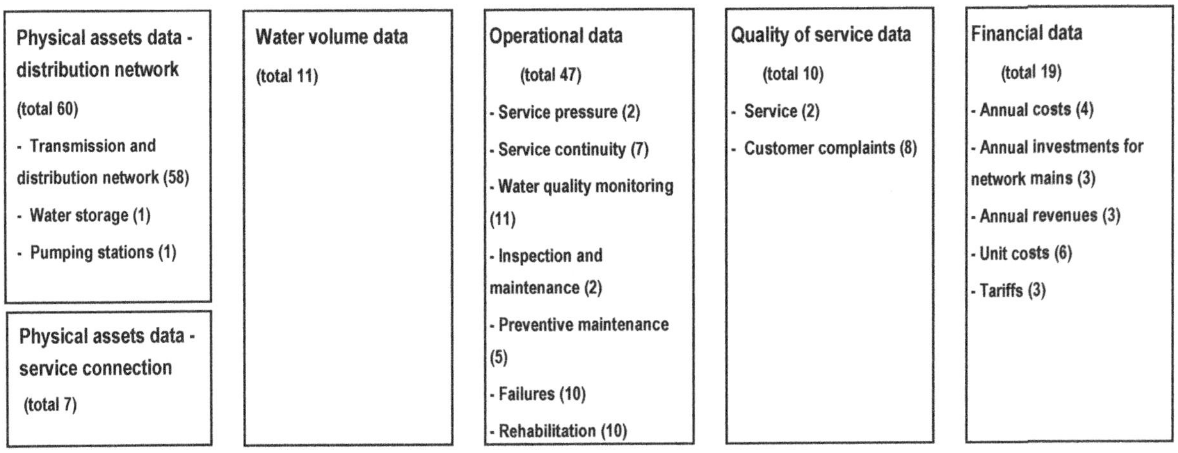

Figure 1.2. Groups of utility information (UI) for network rehabilitation

1.2.4 External information

The EI variables are needed for the CARE-W diagnosis and decision-making process. The users select the adequate EI for the interpretation of each PI, they are grouped in three sections:

- environmental factors (annual rainfall, air temperature, topography);
- mains aggressive factors (physical and chemical soil and groundwater characteristics; geo-technical conditions, traffic class, interference with other infrastructures, seismic conditions);
- economic factors.

The users shall select the EI they consider to be relevant for the interpretation of each PI. The CARE-W PI system includes the following EI:

DESIGNATION	Unit
ENVIRONMENTAL FACTORS	
Annual rainfall:	
average	(mm/year)
maximum	(mm/year)
minimum	(mm/year)
Air temperature:	
daily average	(ºC)
daily maximum	(ºC)
daily minimum	(ºC)
Topography:	
source average elevation	(m)
maximum delivery elevation	(m)
minimum delivery elevation	(m)
MAINS AGRESSIVE FACTORS	
Physical and chemical soil and groundwater characteristics:	
hydrogenionic concentration range (pH)	(pH-pH)
agressive carbon dioxide concentration range (CO_2)	(mg/l-mg/l)
sulphate concentration range (SO_4^{2-})	(mg/l-mg/l)
chloride concentration range (Cl^-)	(mg/l-mg/l)
resistivity concentration range	(Ωm -Ωm)
organic compounds	(yes/no)
stray currents	(yes/no)
Geotechnical conditions:	
high seat stability	(yes/no)
medium seat stability	(yes/no)
low seat stability	(yes/no)
Traffic class:	
heavy traffic	(yes/no)

normal traffic	(yes/no)
light traffic	(yes/no)
Interference with other infrastructures:	
high risk to be affected by other infrastructures works	(yes/no)
medium risk to be affected by other infrastructures works	(yes/no)
low risk to be affected by other infrastructures works	(yes/no)
Seismic conditions:	
range of forecasted maximum soil movement due to liquefaction	(mm-mm)
range of forecasted maximum angular deflection in joints	(º-º)
range of forecasted maximum axial displacement in joints	(mm-mm)
ECONOMIC FACTORS	
Inflation rate	(%/year)

1.2.5 Additional performance measures

As mentioned before, additionally to this set of PI it was decided to create a group of *additional performance measures*, which can be useful for the CARE-W project but are more difficult to be defined, are difficult to assess, do not guarantee to be unbiased and whose thresholds are difficult to establish.

These measures are to be developed by each CARE-W partner according to their own needs and based on their specific tools, such as hydraulic modeling or statistical analysis. Some potential examples of additional performance measures are *network hydraulic reliability*, *mains residual service life*, *service connections residual service life*, *balance of costs and benefits* and *internal rate of return*.

1.3 THE CARE-W PI TOOL

1.3.1 What is the PI Tool?

The PI Tool is a computer program that makes available to the CARE-W water supply network rehabilitation procedure an information system for the calculation, storage, update and retrieval of performance indicators and their associated information (Praça *et al.*, 2002). Apart from the values of indicators and their calculation data, the system provides complete information on all their features, properties and relationships. The PI Tool makes use of the common CARE-W database standard.

The PI Tool is published either as a stand-alone program or as an integrated facility within the CARE-W software package.

1.3.2 Selection of the PI and UI with the PI tool

Most users of the PI tool will probably be interested in a subset of rehab PI, and not in all of them. The PI and UI selection procedures are therefore a key aspect of the PI tool. The **PI/UI Manager** (Figure 1.3) is in many ways the centerpiece of the PI Tool. The **PI/UI Manager** is both a selection facility and an information centre about the PI and UI.

PI and UI have different property sheets (Figure 1.3 shows the UI property sheet). In order to avoid mistaking the two, the UI property sheet uses a light-green background, while the PI property sheet has a light-blue background. The droplet icons used in the selection trees follow the same colour coding.

The droplet icons have another function: a fully-coloured droplet means that the database already contains data for that PI or UI, for the particular dataset and period under analysis; conversely, an "empty" droplet means that there is no data in the database for that PI or UI.

1.3.3 Data input

Data input in PI Tool means typing in EI and UI values through specific input tabular forms. The latter is dependent on PI and/or UI selection having been made through the PI/UI Manager. The procedure used by the program for either EI or UI input forms is to prompt the user for database update on exiting the form. The following sections explain both cases in detail.

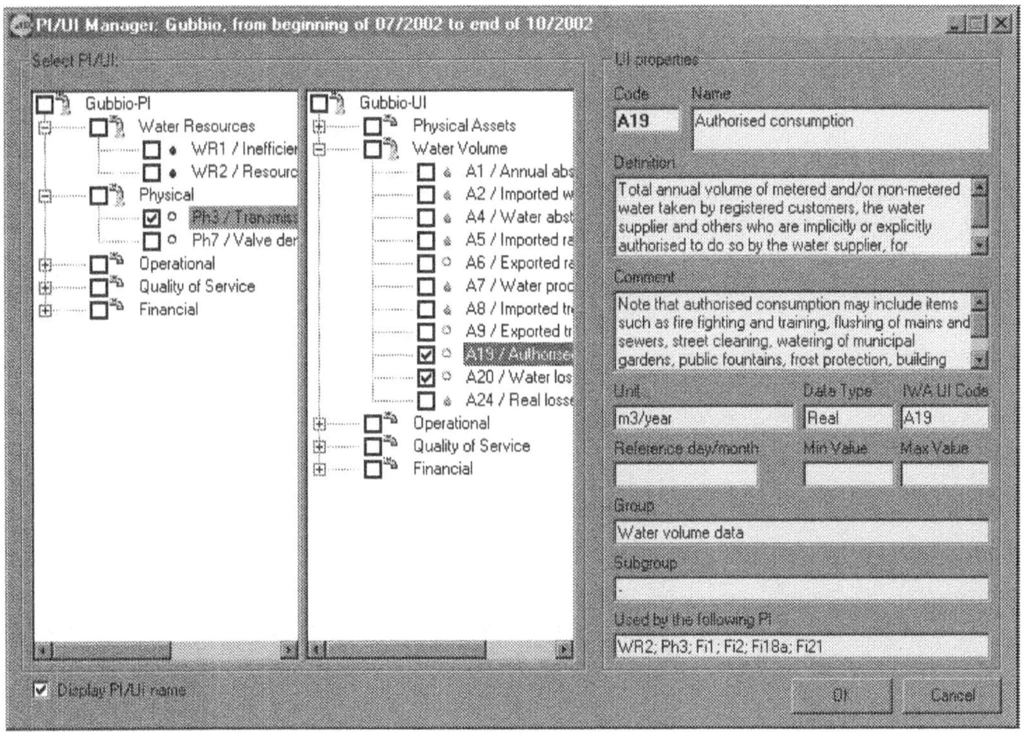

Figure 1.3. The PI/UI Manager, showing the UI property sheet

1.3.4 EI input

External information (EI) helps to provide a context for the performance indicators analysis. The EI for a particular dataset and period are input to the database through the **EI Input** dialog box (Figure 1.4).

The EI are not used when calculating PI. The user may input EI values to the database at any moment during a session, provided a dataset and period have been selected.

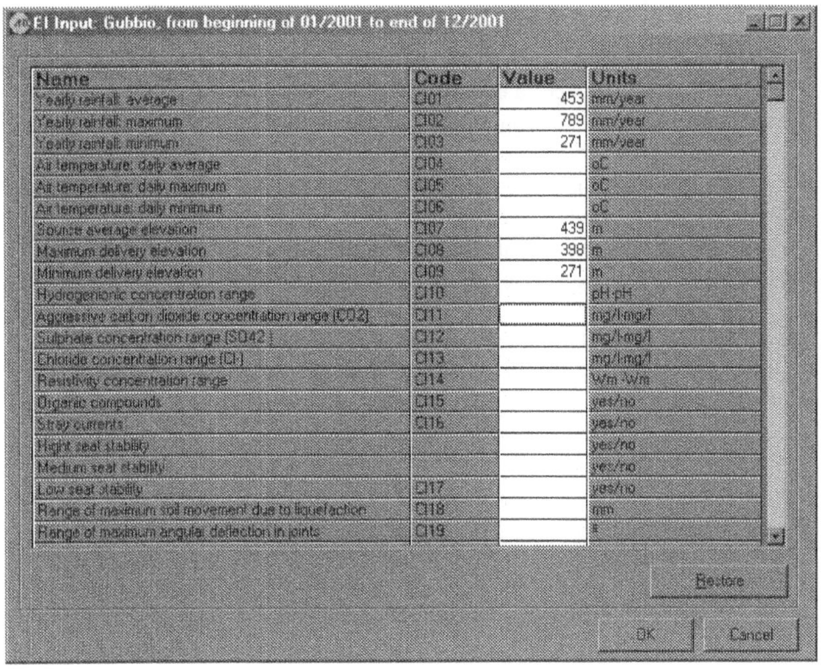

Figure 1.4. The EI Input dialog box

If the database contains values for any EI, for the particular dataset and time period, then the values are loaded when the form appears. The user may modify them at will, as well as type in values for any other EI.

1.3.5 UI input

For a set of PI to be calculated, the database must contain the values for the necessary UI. The place where the user types in the UI values and saves them to the database is the **UI Input** dialog box (Figure 1.5).

As with the EI, if the database contains values for any of the listed UI, for the particular dataset and time period, then the values are loaded when the form appears. The user may modify them at will, as well as type in values for any other UI on the list.

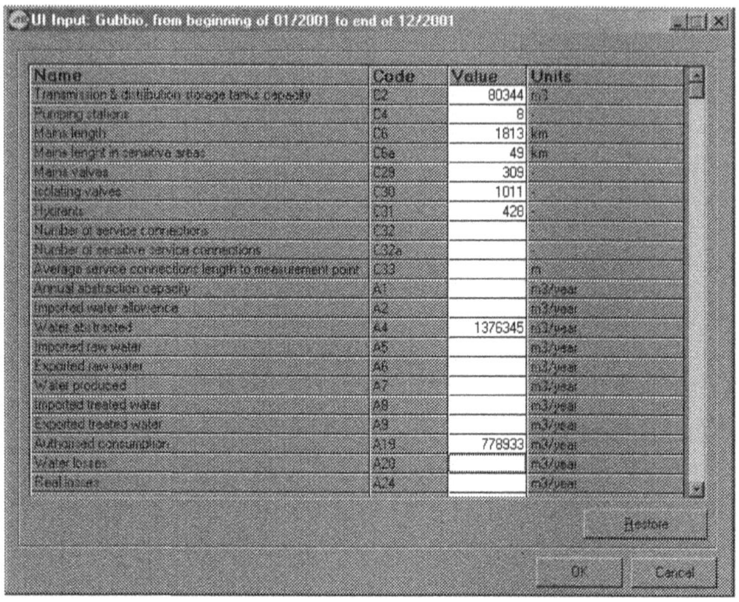

Figure 1.5. The UI Input dialog box

1.3.6 Calculating PI and looking at results

The PI that have been selected in the **PI/UI Manager** are calculated by applying the respective processing rules to the UI values that are stored in the database. The program produces a calculation report that lists:

- the total number of PI that were successfully calculated from the UI values stored in the database for the specified dataset and period;
- the total number and the codes of the PI that were not calculated due to missing UI values, i.e., due the database not containing the UI needed, for the specified dataset and period;
- the total number and the codes of the PI that were not calculated due to division-by-zero errors being generated by stored UI values used in their calculation, for the specified dataset and period.

The program will attempt to calculate even if there are no UI values in the database, but the user will be notified of the unsuccessful results through the calculation report. The program then displays a table containing the PI calculated (Figure 1.6).

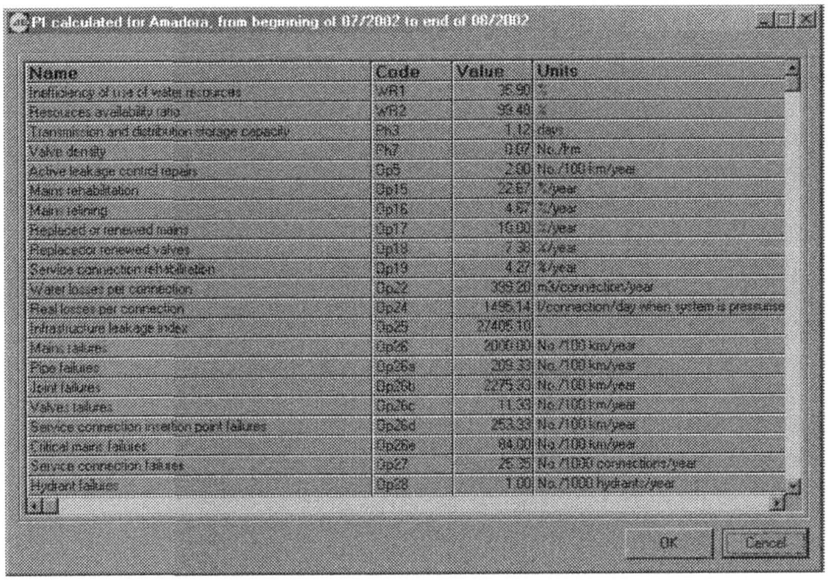

PI calculated for Amadora, from beginning of 07/2002 to end of 08/2002

Name	Code	Value	Units
Inefficiency of use of water resources	WR1	35.90	%
Resources availability ratio	WR2	99.48	%
Transmission and distribution storage capacity	Ph3	1.12	days
Valve density	Ph7	0.07	No./km
Active leakage control repairs	Op5	2.00	No./100 km/year
Mains rehabilitation	Op15	22.67	%/year
Mains relining	Op16	4.67	%/year
Replaced or renewed mains	Op17	18.00	%/year
Replacedor renewed valves	Op18	7.38	%/year
Service connection rehabilitation	Op19	4.27	%/year
Water losses per connection	Op22	399.20	m3/connection/year
Real losses per connection	Op24	1495.14	l/connection/day when system is pressurise
Infrastructure leakage index	Op25	27405.10	-
Mains failures	Op26	2000.00	No./100 km/year
Pipe failures	Op26a	209.33	No./100 km/year
Joint failures	Op26b	2275.33	No./100 km/year
Valves failures	Op26c	11.33	No./100 km/year
Service connection insertion point failures	Op26d	253.33	No./100 km/year
Critical mains failures	Op26e	84.00	No./100 km/year
Service connection failures	Op27	25.35	No./1000 connections/year
Hydrant failures	Op28	1.00	No./1000 hydrants/year

Figure 1.6. The calculated PI

1.3.7 Graphical representations

The Figures 1.7 and 1.8 below give examples of PI presentations. Figure 1.7 is an example of a comparison of authorized water consumption between time periods, and reveals a trend over four years. Figure 1.8 shows an example of a time series graph for D18- Mains rehabilitation generated with the PI tool. It demonstrates variations on mains rehabilitation activities over three years. Analysis of trends for resources spent, condition and performance and comparison between these can give powerful information for the rehabilitation planning process.

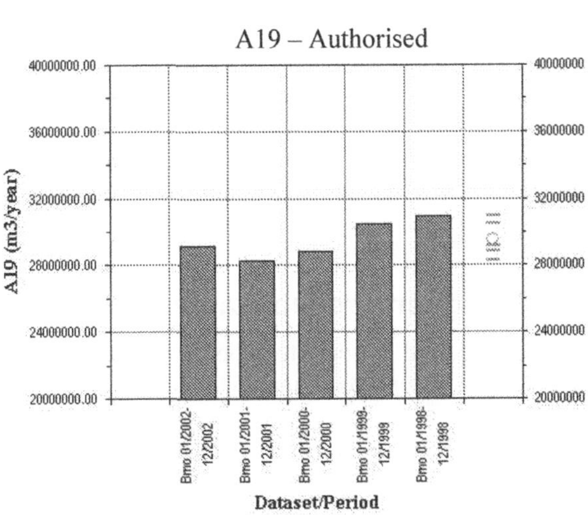

Figure 1.7. Dataset comparison graph of A19 - *Authorised consumption* generated by the PI Tool.

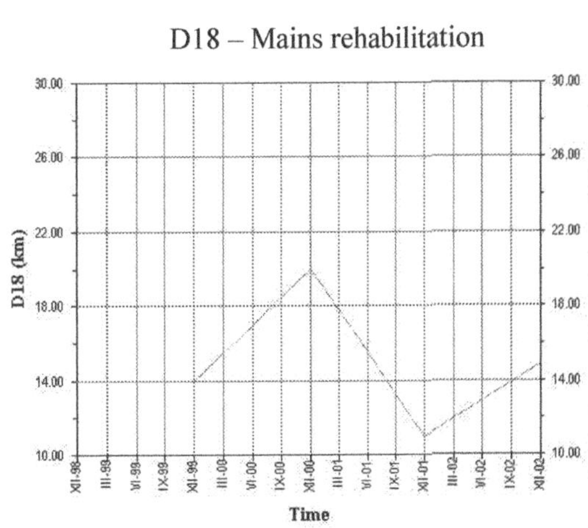

Figure 1.8 Time series graph of D18 – Mains rehabilitation generated by the PI Tool.

1.3.8 The uses of PI within CARE-W framework

It was previously referred that the PI tool can be used as a stand-alone tool, to assist the engineers in analysing the evolution of performance of their networks, in establishing diagnosis or in improving their know-how about the behaviour of different pipe materials or rehabilitation techniques. However, the main objective of the CARE-W rehab PI system is to be used as part of a more comprehensive tool, in close co-ordination with the other modules of the CARE-W system. This section aims to illustrate how this can be done with the example of the use of PI for long-term rehabilitation planning (Baptista & Alegre, 2002 a).

1.3.9 Underlying assumptions for the case of long-term planning

A key objective of long-term rehabilitation planning is to predict the performance of a given water supply network at a future reference time when a rehab scenario is implemented, as well as the difference in performance between this scenario and the *do nothing scenario* (Figure 1.9).

The long-term rehabilitation planning should take into account the *control factors* that the decision makers can play with, and predict the *impact* on the system derived from the implementation of such decisions in order to allow for an easy comparison of the results of different options.

The core *control factors* for long-term planning are the type and rate of the rehabilitation measures. Type includes mains relining, renewed or replaced mains, renewed or replaced valves and service connection rehabilitation. Rate is typically expressed in terms of percentage of the total mains length per year.

Impacts can be quantified in the form of the performance assessment of the system, technically and economically measured from the viewpoints of the physical assets, of the system operations and of the quality of the service provided. Not only the final performance derived from a given decision should be predicted, but also the performance variation between this situation and the *do nothing* scenario is relevant.

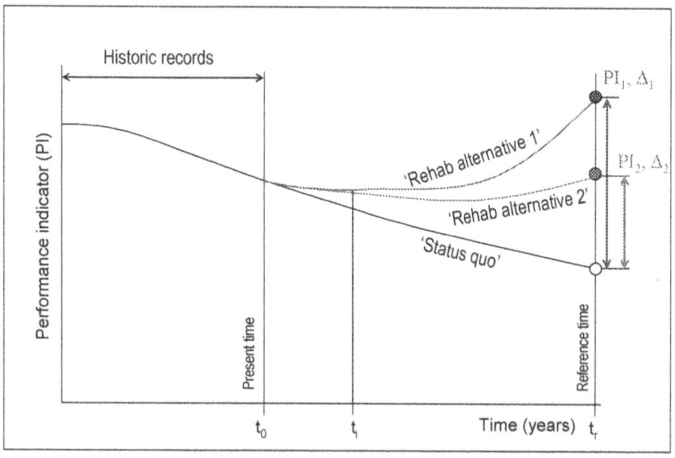

Figure 1.9. Variables to be predicted for long-term rehab planning

Prediction of *impacts* can be based on *explanatory factors* that affect the system performance. Some of these factors are controllable by the undertaking (e.g. service pressure, preventive maintenance practices) while other are not (e.g. soil characteristics, temperature).

1.3.9 Proposed methodology for the use of PI for long-term rehabilitation planning

Based on these underling assumptions, the CARE-W PI system may be used not only to support the diagnosis and benchmarking but also to assist long-term rehabilitation planning. The proposed methodology comprises the following steps:

Step 1: Identification of the *control factors* based on the rehab PI referred above (Figure 1.10), according to the type of rehabilitation selected:

- *Mains rehabilitation (%/year): Length of transmission and distribution mains rehabilitated during the year / total mains length x 100*
  - *mains relining (%/year)*
  - *replaced or renewed mains (%/year)*
  - *replaced or renewed valves (%/year)*
- *Service connection rehabilitation (%/year): Number of service connections replaced or renewed during the year / total number of service connections x 100*

Step 2: Identification of predictable PI for assessing *impacts* of rehabilitation (Figure 1.9 and Figure 1.10), based on the rehab PI referred above, selected according to the type of rehabilitation:

Physical indicators:

- *Mains residual life (years)*

- *Service connections residual life (years)*

Operational indicators:

- *Mains failures (No./ 100 km/ year)*: Number of mains failures during the year, including failures of pipes, valves, fittings and service connection insertion point failures / total mains length x 100
- *Service connection failures (No./ 1000 connections/year)*: Number of service connection failures during the year / number of service connections x 1000
- *Real losses (l/ connection/ day)*: Real losses x 1000 / (number of service connections x 365 x T/100) (T = % of year system is pressurised)
- *Infrastructure leakage index (-)*: Real losses/ technical achievable low-level annual real losses (when system is pressurised)

Quality of service indicators:

- *Pressure of supply adequacy (%)*: Number of delivery points that receive and are likely to receive pressure equal to or above the guaranteed or declared target level at the peak demand hour (but not when demand is abnormal) / number service connections x 100
- *Water interruptions (%)*: Σ (Population subject to a water interruption x duration of the interruption in hours) / (population served x 24 x 365) x 100
- *Water taste test compliance (%)*: Number of water taste tests of treated water complying with the applicable standards or legislation during the year / total number of water taste tests of treated water performed during the year x 100
- *Water colour test compliance (%)*: Number of water colour tests of treated water complying with the applicable standards or legislation during the year / total number of water colour tests of treated water performed during the year x 100

Step 3: Prediction of the curve between the present time and the reference time for the 'do nothing scenario' (Figure 1.10), for each of the previously selected PI, adopting a time series analysis to the historic records and extrapolating the results for the future, after a consistency analysis of the historic records.

Step 4: Prediction of the curve between the present time and the reference time for the 'rehabilitation scenario' (Figure 1.9), for each of previously selected PI, adopting specific prediction functions that take into account the relevant key 'explanatory factors' in the form of curve parameters or coefficients.

Step 5: Impact assessment for each of the previously selected PI, and quantification of the performance variation between the 'rehabilitation scenario' and the 'do nothing scenario' at the reference time (Figure 1.9).

Step 6: Economical analysis (Figure 1.10) between the rehabilitation costs (resulting from Step 1) and the impact assessment (resulting from Step 5) regarding investment delay on new mains (physical indicators), operational cost minimization regarding reduction of water losses and repairs (operational indicators) and social cost minimization for customers regarding pressure, interruptions and water quality (quality of service indicators).

It is then possible to decide which scenario is the more convenient for the long-term rehabilitation planning.

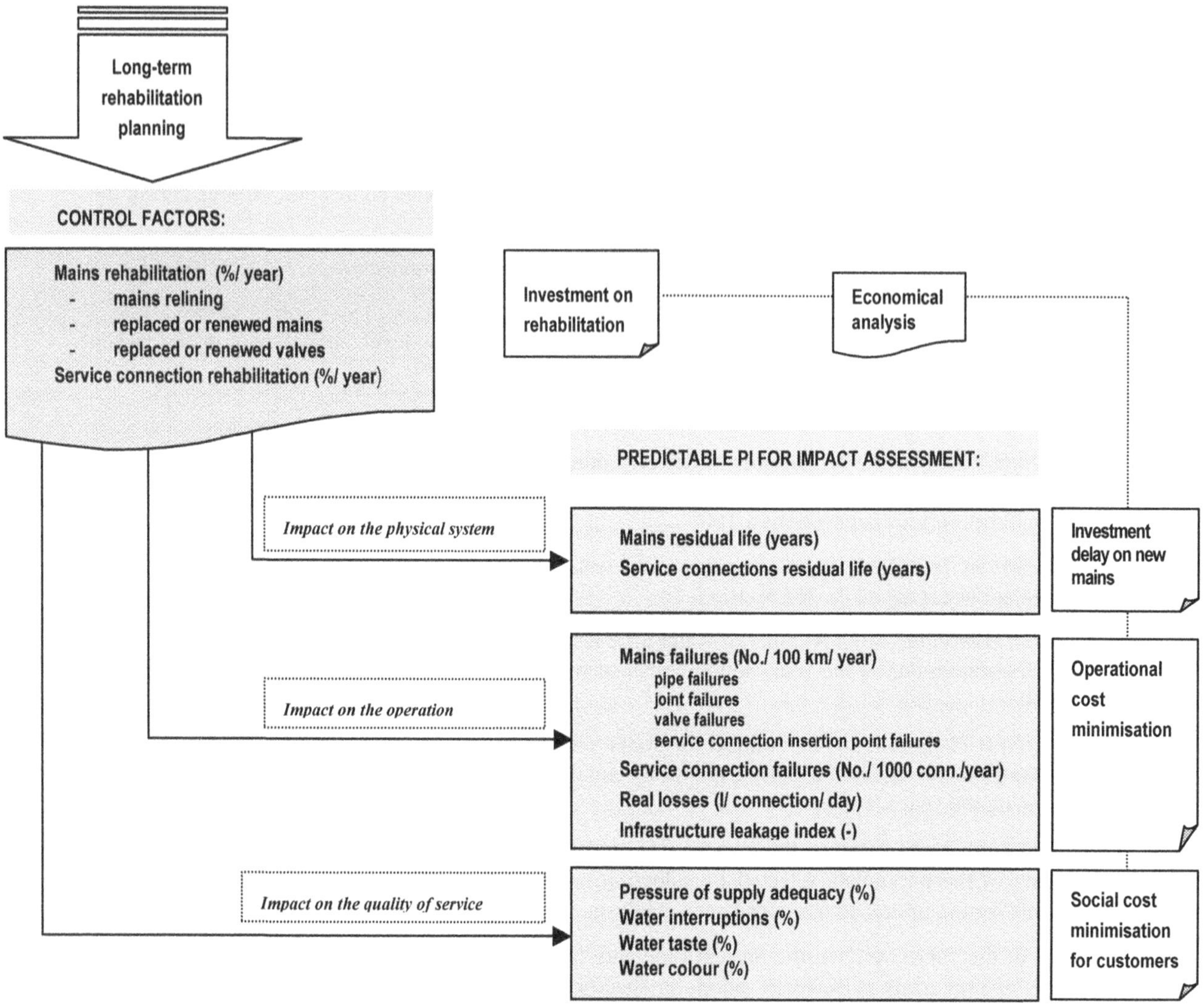

Figure 1.10. CARE-W PI system for long-term rehabilitation planning

1.4 REHAB PI COMPARISON AMONG CARE-W END-USERS

1.4.1 Participants in the international comparison of rehab PIs

All the CARE-W end-users provided data for the establishment of a PI database in close collaboration and with the assistance of the CARE-W corresponding partners (Alegre *et al.*, 2003). Other participations could be included as well, from Czech Republic (Water Supply Joint Stock Company - VAS), France (Roubaix Tourcoing) and Portugal (Loures and Sintra), Table 1.2 summarises the contributions received (totalling 35 case studies).

Table 1.2. Participants in the rehab PI comparison

Partner	Case studies	PI, UI and EI	
		Scope	Years
Bologna University	Massafiscaglia, Codigoro, Italy	Sector	A single set including 1995-2000
	CADF, Codigoro, Italy	Whole network	1995-1996-1997-1998-1999-2000
	Vigarano-Mainarda, Ferrara, Italy	Sector of ACOSEA (Ferrara)	2001
	Reggio Emilia, Italy	Whole network	1996-1997-1998-1999-2000
Brno University	BVK - Brno, Czech Republic	Whole network	1998-1999-2000-2001
	Water Supply Joint Stock Company (VAS a.s.) Czech Republic	Whole network + 6 sectors	2001
CEMAGREF	Roubaix/Tourcoing, France	Whole network	2001
	Lausanne, Switzerland	Whole network	2001
Dresden University	Drewag – Dresden, Germany	Whole network	2001
	Stuttgart, Germany	Whole network	2001
LNEC	Oeiras & Amadora	Amadora network	2001
	Sintra	Whole network	2002
	Loures	Whole network	2001
INSA	Lyon CGE, France	Whole network	2001
SINTEF	Trondheim, Norway	Whole network	2001
	Oslo, Norway	Whole network	2001
WRc	Bristol, United Kingdom	1 water supply zone	2001

1.4.2 Results achieved

About a half of the indicators were assessed by at least one half of the case studies, as shown in Figure 11, and 19 of them were assessed by 21.

The methodology adopted for data processing included (i) data validation, (ii) codification of the datasets, (iii) data input and PI reassessment, (iv) statistical analysis and graphical representation of the results, (v) identification of the difficulties of assessment, (vi) identification of the relevance of the PIs, (vii) preliminary definition of guidance ranges, and (viii) assessment of global statistics and elaboration of the individual PI forms.

The global statistics are presented in the form of box & whisker plots, for all the indicators for which at least 7 replies were received. These plots show the quartiles (0, 25%, 50%/median, 75% and 100%) and the mean value (marked as a filled dot, ●). In some cases outliers and extreme values occurred. An outlier is a value of data that is more than 1.5 times the interquartile range from the quartiles and is plotted as O. An extreme value is a value of data that is more than 3 times the interquartile range from the quartiles and is plotted as *.
The representation of results in this form combined with the indication of the guidance range provides in a rather synthetic way a broad overview of results achieved.

Figure 1.11 exemplifies this for one of the groups of rehab PI, the 'operational indicators' group.

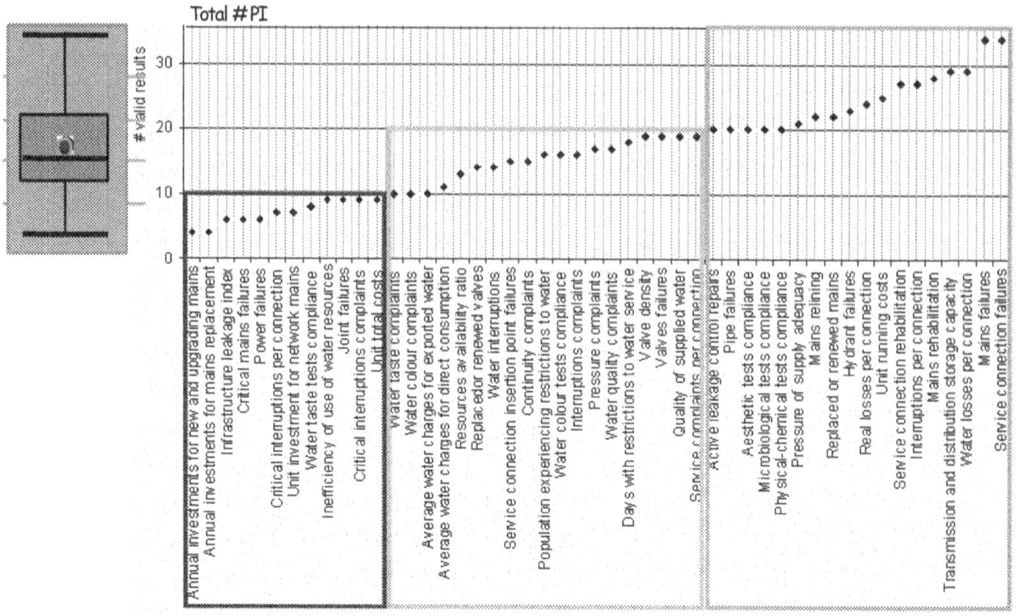

Figure 1.11. Total number of valid results per performance indicator (PI)

1.4.3 Preliminary definition of guidance ranges

One of the aims of the test was the establishment of guidance ranges for each rehab PI, so as the users can mores easily interpret their own results. Guidance ranges are not meant to judge the results, as it is frequent that a given indicator has an "abnormal" value for a well know and fully acceptable reason, not because the undertaking has a poor performance. For instance, the guidance range for the PI *Op15 - Mains rehabilitation (%/year)* was established between 1 and 2.0% of mains rehabilitated per year representing a global network renovation within a period of 50 to 100 years, assuming that this is the average life time range for the most commonly used pipe materials. However, for instance for network recently built, it is logic that the rehabilitation rate is much lower. Every case needs therefore to be analysed individually before a judgment is established and a decision made from it.

The guidance ranges proposed do not correspond to the statistics of the test made. The sample is not sufficiently big and representative to provide valid values. Conversely, results might be biased because there are some cases with several sectors from the same network, and other cases for which there are several years of data for the same network.

Several sources and justifications were adopted, depending on case to case. For some of the indicators, there are already values referenced in the literature that can support the establishment of guidance ranges, and were adopted whenever possible. However, this is not the case for most of them. If a clear source or background reason is not presented in the PI form, this means that the authors made a preliminary estimate based on their own professional experience, but not on a sound basis. All end-users received a preliminary version of the results and were invited to comment these values. Some contributions were received and taken into account, particularly from Trondheim and from SINTEF.

This is an evolving process, and the application of the rehab PI's to other case studies will allow for collecting more data and consolidating the existing knowledge.

1.4.4 Assessment of global statistics and elaboration of the individual PI forms

The final stage of data processing consisted of the assessment of the global statistics the elaboration of individual PI forms.

The global statistics are presented in the form of box & whiskers plots, for all the indicators for which at least 7 replies were received. In some cases outliers and extreme values occurred. An outlier is a value of data that is more than 1.5 times the interquartile range from the quartiles and is plot as O. An extreme value is a value of data that is more than 3.0 times the interquartile range from the quartiles and is plot as *. The mean value is represented as •. The quartiles are represented as shown in Figure 1.12.

Figure 1.12 – Representation of quartiles in the box & whiskers plots

Two sets of results are presented. For the former set (graphs with a yellow border), the historical data has been limited to one period of time per network (or sector) to prevent bias caused by a greater weight of the networks for which there are several years of data. For the latter (graphs with a red dashed border) the whole sample has been included.

The individual PI forms synthesise the results, and include the definition of the PI (code, name, unit, concept, processing rules and listing of variables used), a graphical representation of the values collected, the guidance range proposed (whenever possible) and a short explanation on how it was established, some comments on the results, a description of the main assessment difficulties reported by the participants and the relevance of the indicator from the participants viewpoint. The graphical representation includes also a box & whiskers plot for the whole sample, without making any differentiation between case studies (e.g. type, date).

Figure 1.13 exemplifies one of these forms and explains their contents.

Op15 - **MAINS REHABILITATION** (%/year)

Concept: Length of transmission and distribution mains rehabilitated during the year / total mains len

Processing rule: Op15 = D18/C6 x 100

Variables: D18 – Mains rehabilitation (km/year)
C6 – Mains length (km)

PI value for each case study

Box & whiskers:
- max
- 75%,
- median
- 25%
- min
- average

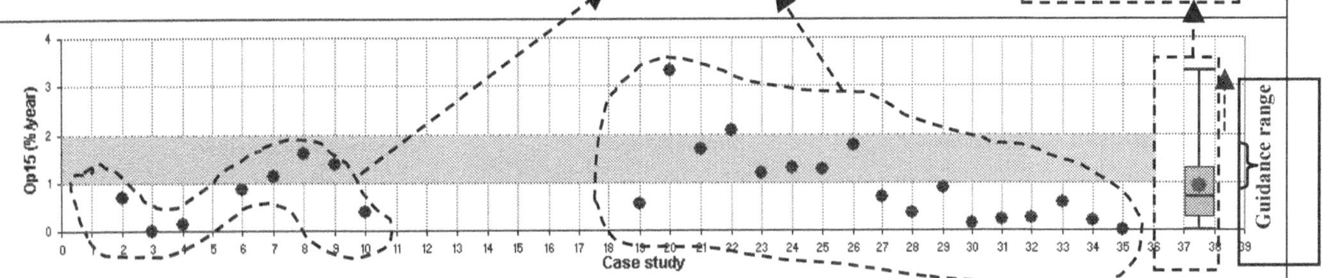

Comments:	The mean value of the results collected is 0.91% of mains rehabilitated per year and 50% of the case studies are situated in the range from 0.25 to 1.3% of mains rehabilitated per year. As it might be expected, a significant number of case studies present relatively low rehabilitation rates, corresponding in average to a mains lifetime of 110 years.
Difficulties:	A few end-users had difficulties with D18 (Mains rehabilitation). This difficulty seems to be due to the lack of record, not to the complexity of the variable itself, and nobody recommended abandoning this PI due to assessment difficulties.
Relevance:	This PI is considered essential or important by all end-users.
PI guidance range:	At this stage, a tentative recommended guidance range seems to be 1.0-2.0% of mains rehabilitated per year, representing a global network renovation within a period of 50 to 100 years.

Figure 1.13. Example (with complementary legends) of a PI form

Section 0 presents the global statistics. The individual PI forms are presented in Appendix 2.

1.5 RESULTS ACHIEVED

1.5.1 Operational indicators

In both the case of the “yellow” set (corresponding to one element per network or sector) and of the red set (whole sample) all the operational rehab PI’s are represented, excepted for Op26b, Op26e, Op29 and Op25, for which there are insufficient data (Figure 1.14).

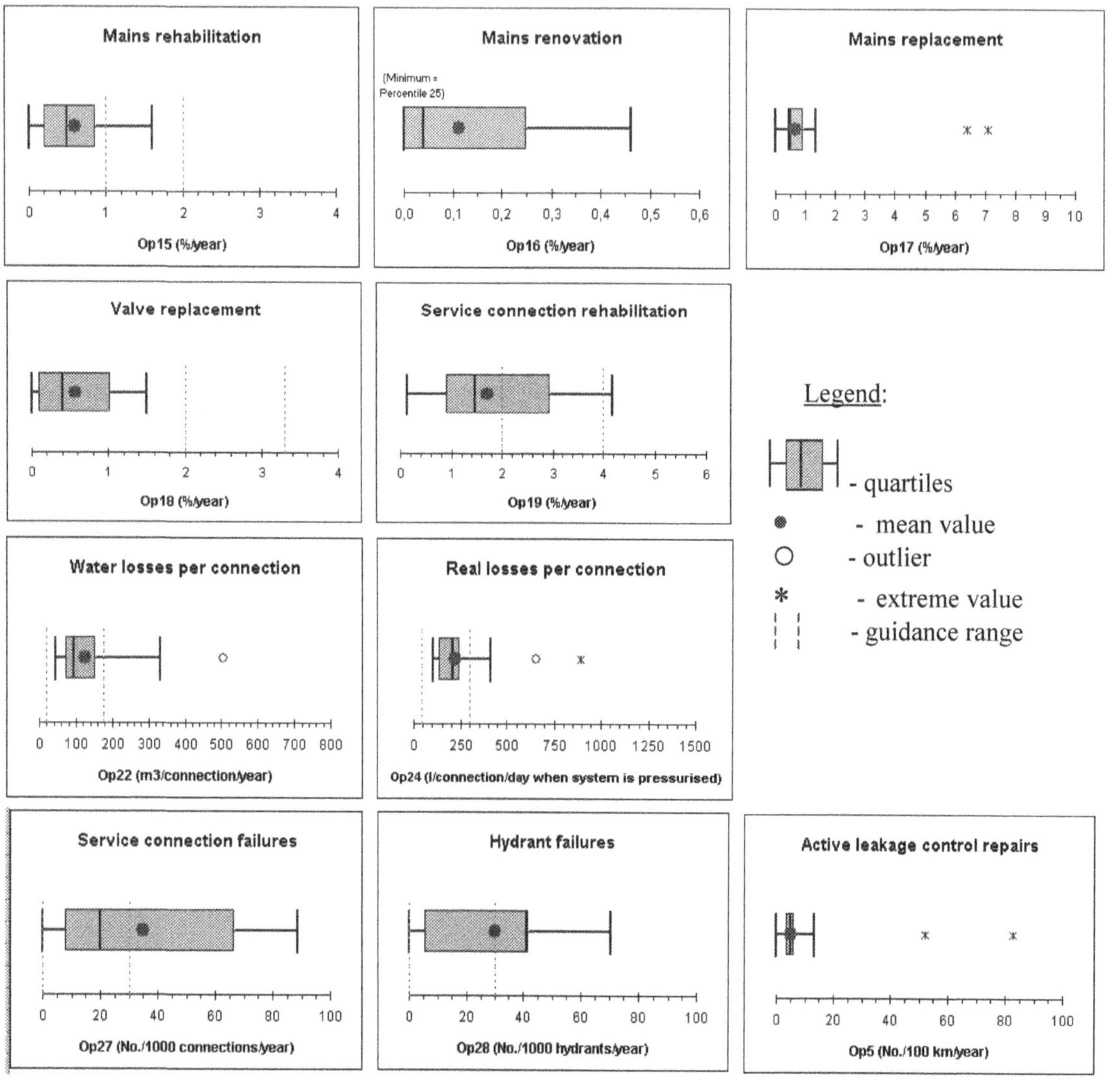

Figure 1.14. Example of results synthesised in box and whisker plots for the PI group ‘Operational Indicators’

1.5.2 Quality of service indicators

In the case of the “yellow” set (corresponding to one element per network or sector), the indicators QS11, QS12a, QS16a, QS24, QS25, QS25a, QS25b, QS26 and QS26a could not be assessed. For the “red set” (whole sample), only the indicators QS12a, QS16a, QS26a could not be assessed (figures 1.15, 1.16).

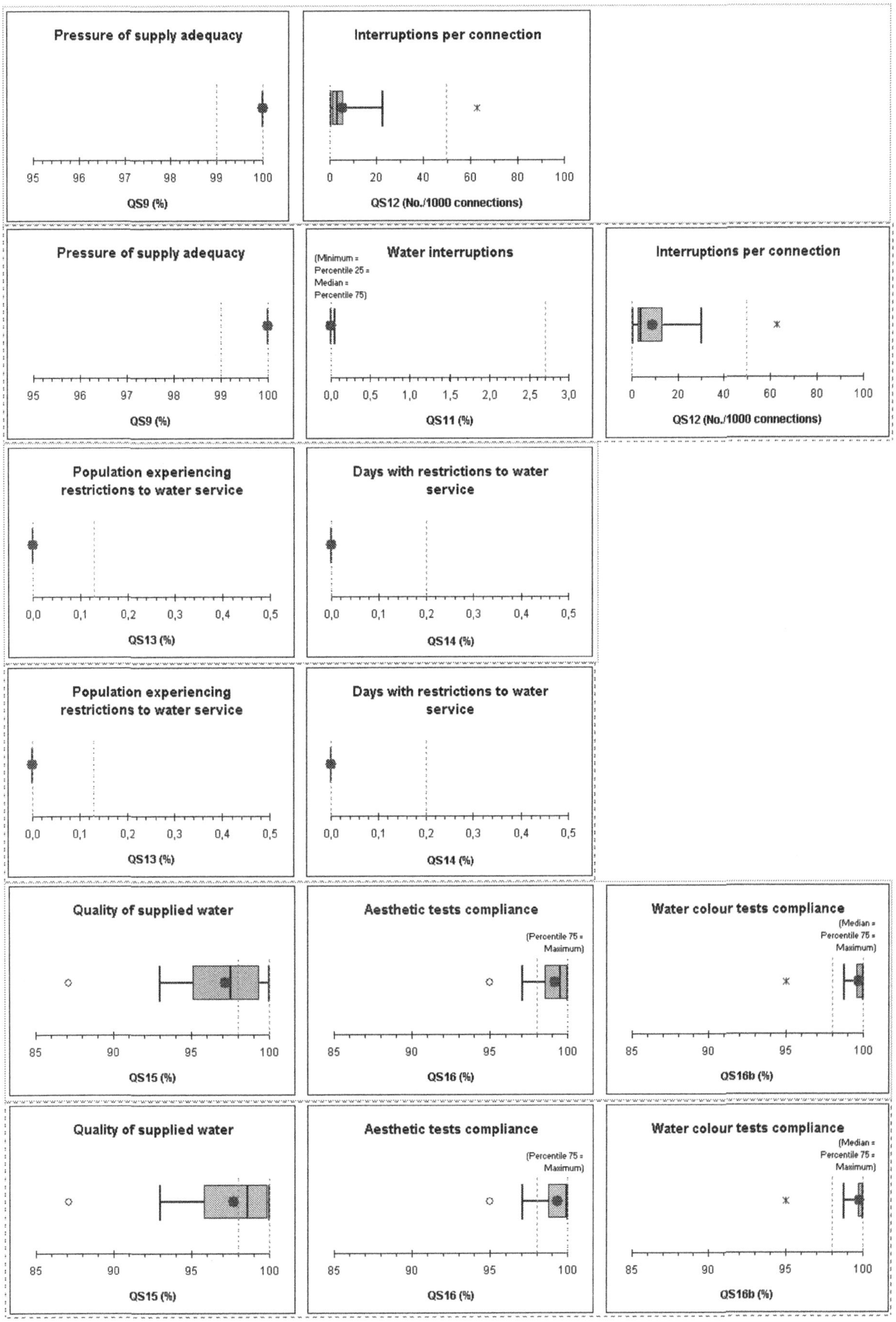

Figure 1.15. Quality of service (1/2)

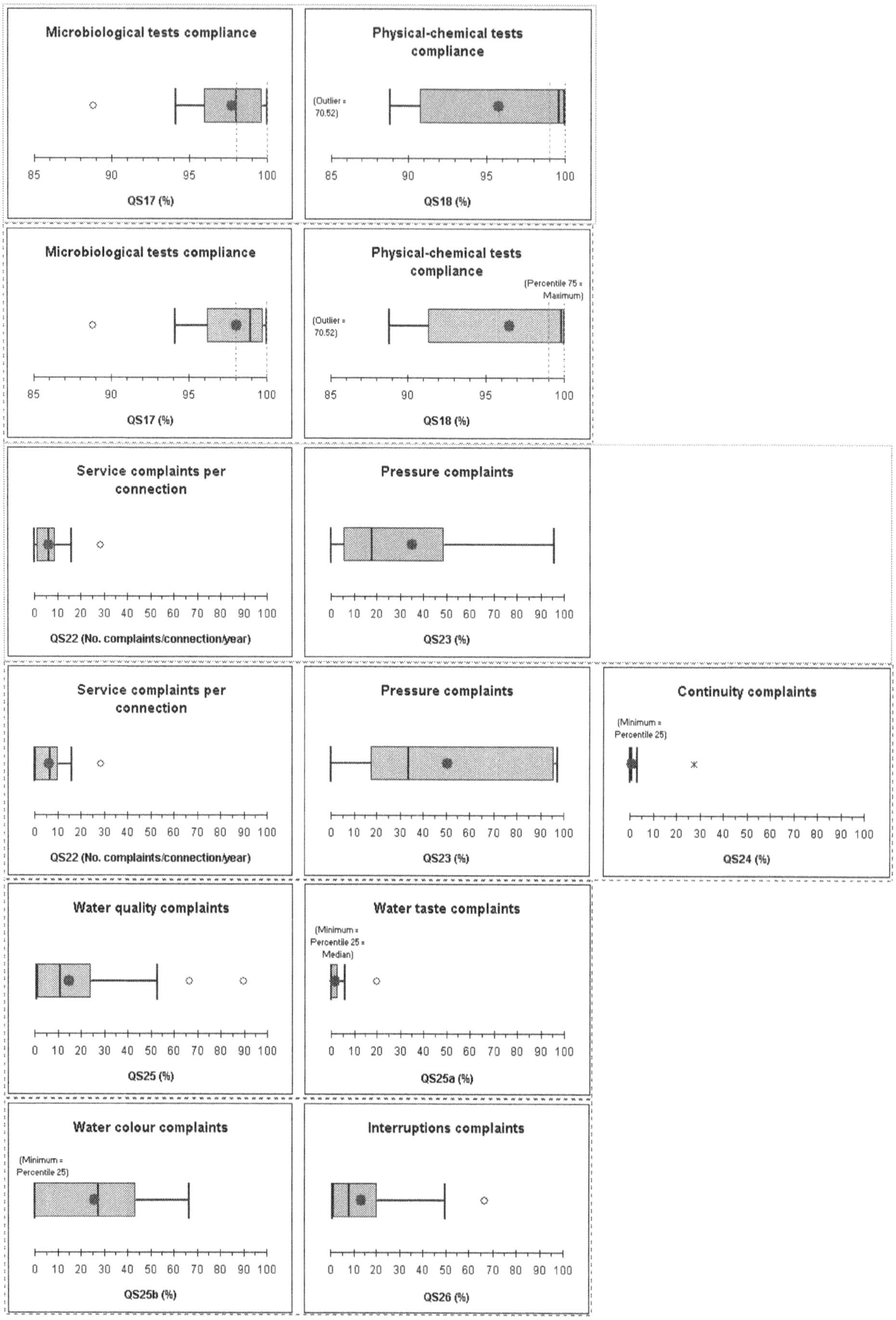

Figure 1.16. Quality of service (2/2)

1.5.3 Financial Indicators

The financial indicators Fi18a, Fi19a, Fi20a and Fi22 could not be assessed in none of the sets. In case of the "yellow set", Fi21 could not be assessed either (Figure 1.17).

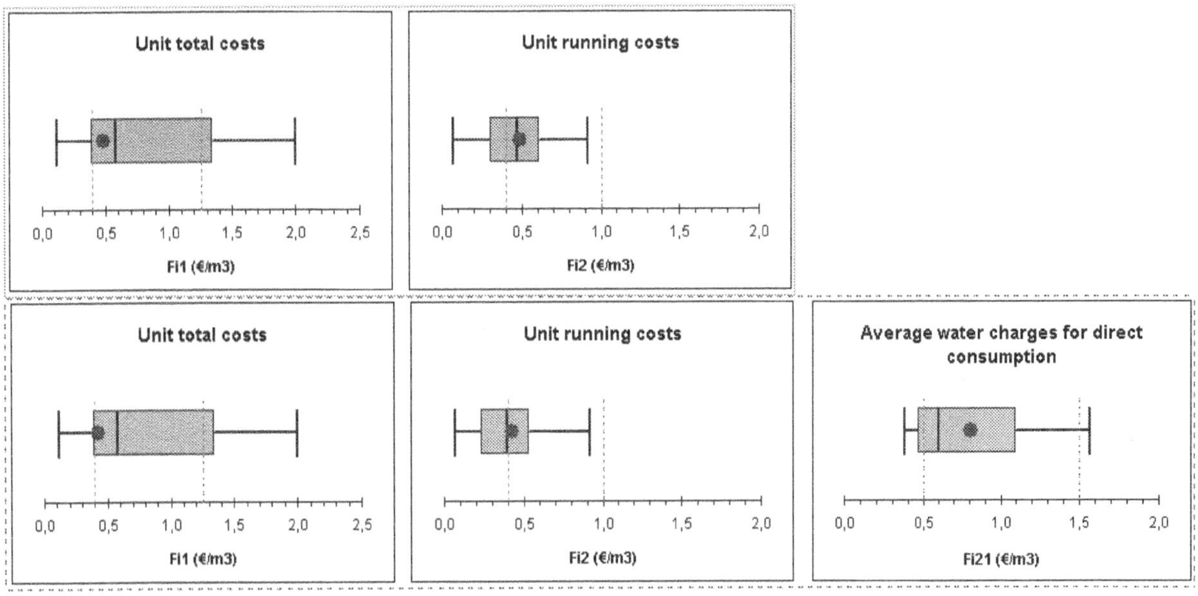

Figure 1.17. Financial indicators

1.5.4 Water resources indicators

Both water resources indicators could be assessed for the red set, but WR1 could not in the case of the "yellow set" (Figure 1.18).

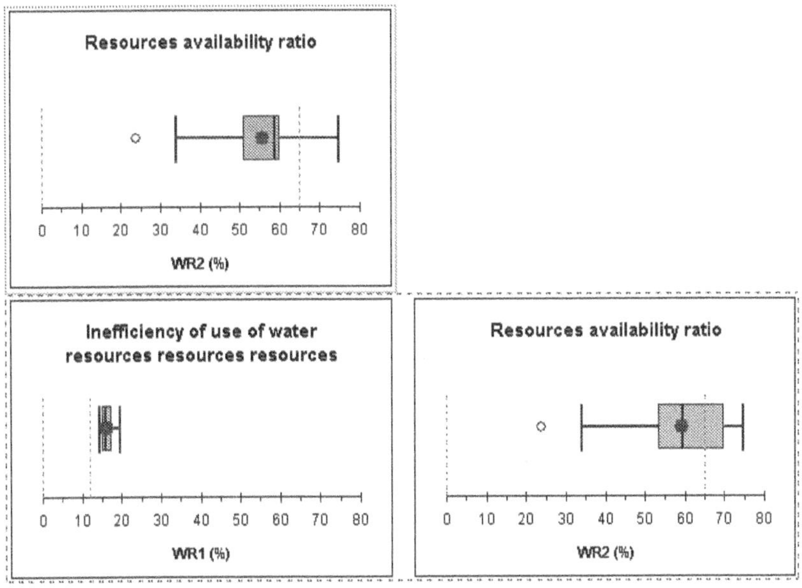

Figure 1.18. Water resources

1.5.5 Physical indicators

The two rehabilitation physical indicators could be assessed for both sets, Figure 1.19.

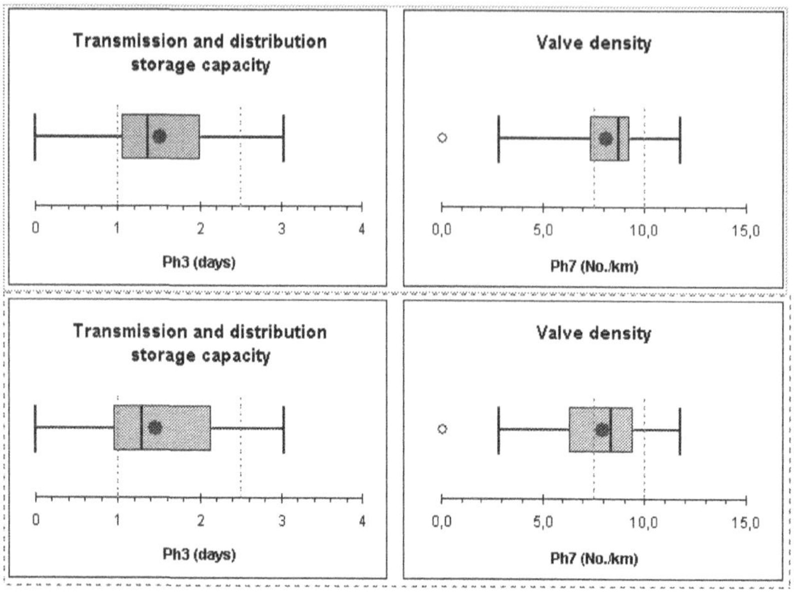

Figure 1.19. Physical indicators

For each PI, an individual form was also elaborated. It includes the definition of the PI (code, name, unit, concept, processing rules and listing of variables used), a graphical representation of the values collected, the guidance range proposed (whenever possible), indicated with the grey filled area, and a short explanation on how it was established, some comments on the results, a description of the main assessment difficulties reported by the participants and the relevance of the indicator from the participants viewpoint. The graphical representation includes also a box & whisker plot for the whole sample. Figure 1.20 exemplifies one of these forms. Appendix 5 presents all the results achieved.

1.6 CONCLUDING REMARKS

Quantitative and objective PI system is a powerful management tool to assist in problem diagnosis, identifying areas for improvement and therefore supporting the development rehabilitation strategies.

The IWA PI system provides a comprehensive reference. The CARE PI system is based on the IWA-PIs. The CARE PI Tool is a dedicated software application integrated in the CARE-W Rehabilitation Manager that allows for PI and UI selection, data input, import and export, PI assessment, result reporting and in graphical and tabular representations.

Comparisons amongst the CARE-W end-users based on the CARE-W rehab PI system has proven to be rewarding and the feedback from the CARE-W end-users is encouraging. Further data collection is needed to get more reliable statistical information and to improve the determination of the upper and lower bounds recommended for each PI.

Op15 - **MAINS REHABILITATION** (%/year)

Concept: Length of transmission and distribution mains rehabilitated during the year / total mains length x 100

Processing rule: Op15 = D18/C6 x 100

Variables: D18 – Mains rehabilitation (km/year) C6 – Mains length (km)

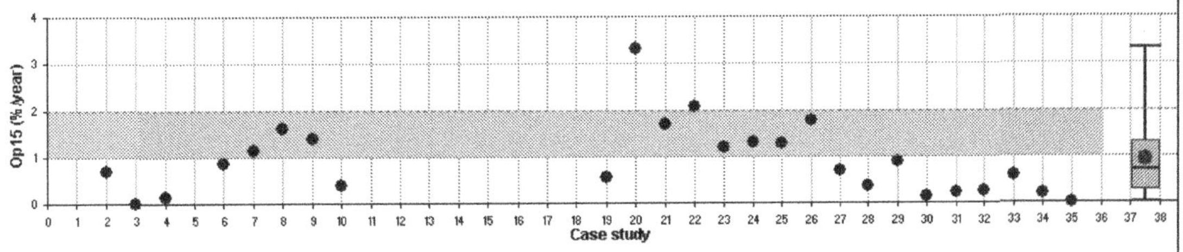

Comments: The mean value of the results collected is 0.91% of mains rehabilitated per year and 50% of the case studies are situated in the range from 0.25 to 1.3% of mains rehabilitated per year. As it might be expected, a significant number of case studies present relatively low rehabilitation rates, corresponding in average to a mains lifetime of 110 years.

Difficulties: A few end-users had difficulties with D18 (Mains rehabilitation). This difficulty seems to be due to the lack of records, not to the complexity of the variable itself, and nobody recommended abandoning this PI due to assessment difficulties.

Relevance: This PI is considered essential or important by all end-users.

PI guidance range: At this stage, a tentative recommended guidance range seems to be 1.0-2.0% of mains rehabilitated per year, representing a global network renovation within a period of 50 to 100 years.

Figure 1.20. Example of a PI form

1.7 REFERENCES

Alegre, H, Baptista, J.M.; Coelho, S. T., Praça, P. (2002). *Performance indicators for network rehabilitation.* International Conference "Computer Aided Rehabilitation of Water Networks CARE-W". Dresden, 1 November (pp. 53-64).

Alegre, H. (2002) - *Performance indicators as a management support tool*, chap. 9, in "Urban Water Supply Infrastructure Management Handbook", ed.-in-chief Larry Mays, McGraw Hill, ISBN 0-07-137160-5 (pp. 9.1-9.74).

Alegre, H.; Baptista; J.M., (2003) – WP1 - *Construction of a control panel of performance indicators for rehabilitation: Rehab PI guidance range.* Report No. 1.5 of the project CARE-W: Computer Aided Rehabilitation of Water networks. 5th Framework Programme of the European Union. LNEC. Lisbon. Portugal (95 pp.) (classified electronic edition, http://www2.unife.it/bscw/bscw.cgi/).

Alegre, H.; Hirner; W.; Baptista, J.M.; Parena, R. (2000). *Performance indicators for water supply services.* Manual of Best Practice Series, IWA Publishing, London, ISBN 1 900222 27 2 (160 pp.).

Baptista, J.M.; Alegre, H. (2002). *Construction of a control panel of performance indicators for rehabilitation: Validation of the rehabilitation PI system.* Report No. 1.2. WP1, CARE-W project, 5th Framework Programme of the European Union, LNEC (100 pp.).

Baptista, J.M.; Alegre; H. (2001). Preliminary proposal of a performance indicator system for rehabilitation, Report No. 1.1. WP1, CARE – W, Computer Aided REhabilitation of Water networks. Decision Support Tools for Sustainable Water Network Management. WP1 - Construction of a control panel of performance indicators for rehabilitation. Lisbon, June 2001.

Baptista, J.M.; Alegre, H. (2002b). WP1-Construction of a control panel of performance indicators for rehabilitation: Definition of a predictable performance indicators for long-term rehabilitation planning. Report No. 1.3 of the project CARE-W: Computer Aided Rehabilitation of Water networks. 5th Framework Programme of the European Union. LNEC. Lisbon. Portugal (23 pp.).

Praça, P.; Coelho, S.T.; Vispi, P. (2002). PI Tool users manual. WP1-Construction of a control panel of performance indicators for rehabilitation Report No. 1.4 of the project CARE-W: Computer Aided Rehabilitation of Water Networks CARE-W project. 5th Framework Programme of the European Union, LNEC. Lisbon. Portugal (38 pp.).

2

Technical tools for failure prediction and water reliability analysis

Patrick Eisenbeis, Cemagref

2.1 BACKGROUND, BASIS FOR THE WORK

Chapter 3 of this book (Decision support for annual rehabilitation) presents the criteria to consider for an Annual Rehabilitation Programme (ARP). Some of these criteria take into account the consequences of water pipes' poor condition on the water utility and the environment. For instance, they include repair cost and the risk of flooding or disturbance to the pipe environment.

The assessment of a large proportion of these criteria requires the knowledge of failure rates, predicted for each pipe. These criteria are listed in Table 2.1. It is then necessary to forecast failure rates in order to classify pipes and make comparisons amongst them.

Table 2.1. Criteria of CARE-W_ARP that require knowledge of failure rates

Criteria	Definition
ARC	Annual Repair Costs
PWI	Predicted Water Interruptions
PCWI	Predicted Critical Water Interruptions
PFWI	Predicted Frequency of Water Interruptions.
DFH	Damages due to Flooding in Housing areas
DFI	Damages due to Flooding in Industrial or Commercial areas
DSM	Damages due to soil movements
DT	Traffic Disruptions
DDI	Damage and/or Disruption on other Infrastructure

Three other technical criteria, assessing condition and importance of water pipes, are also useful for CARE-W_ARP:

- Hydraulic criticality: this criterion combines failure or burst rate with hydraulic importance of the pipe, linked to its location in the network and to the type of network,
- Water losses, caused by the pipe
- Impact of pipe condition on supplied water quality.

With the help of hydraulic models, the first indicator, hydraulic criticality, can be calculated, integrating eventually, pipe failure rate. Water loss assessment on the pipe scale is more difficult. This can be done, at zone level, by dividing the network into sectors and performing step tests or continuous measurement. The indicator can be assessed at zone level by considering customer complaints or using assumptions on old and unlined pipes. Research of the impact of pipe condition on water quality continues. This research effort is focussed in two areas: the assessment of the chlorine decay rate, using hydraulic modelling, and the measurement of iron concentration in corroded pipes.

In CARE-W_ARP, the last two indicators above are assessed by evaluating by pipe group or zone, based on the water utilities' knowledge. However, the indicators "failure rate" and "hydraulic criticality" are also computed at the pipe level using the computation tools presented below.

Two major kind of tools are studied:

- tools, whose objectives are to forecast failures on a pipe or a group of pipes,
- tools, whose objectives are to assess hydraulic reliability of a pipe or a group of pipes.

Originally 8 models were foreseen: 5 forecast failures models and 3 hydraulic reliability models (Cf. Table 2.2). In the tests presented below, only 5 (in "orange") of them have been studied. A precise description of the models is given in the report D3 (Eisenbeis et al, 2002)

Table 2.2. Models studied in the Work Package

	Forecast failure models					Hydraulic reliability models		
CARE-W acronym	PHM model	Markov model	Poisson analysis		NHPP model	Aquarel	Failnet-Reliab	Relnet
Research Centres/ Universities	Cemagref	INSA-Lyon	INSA-Lyon	SINTEF	NTNU	SINTEF	Cemagref	Brno University

2.1.1 The statistical analyses used for the Failure Forecasting models

The Failure Forecasting tools (FAIL tools) are *CARE-W_PHM* (Cemagref) and *CARE-W_Poisson* (INSA-Lyon). They are based on the knowledge of past, observed and recorded failures. A failure is defined in this case as a break or detected leak that has necessitated repair to the pipe.

These forecast tools require the existence of a database, sufficiently accurate, characterising network pipe links and listing failures and their occurrence date.

The two tools are differentiated by their statistical approach:

- *CARE-W_PHM* uses data from previous failures to develop failure forecast models, based on specific methods used in epidemiology called survival analysis. These methods analyse the time between two failures and assess the influence of different risk factors specific to the pipe or to its environment. They lead to the production of failure models, based on the Weibull model and described by the survival and hazard functions. After defining significant variables, predicted failure numbers and rates at 5 or 10-year horizons may be computed for each pipe.
- *CARE-W_Poisson* [(4)] computes mean failure rate by pipe group, where groups are established according to pipe characteristics, such as diameter and material, or the environment, such as soil type and traffic loading. The influence of these factors can also been studied by a statistical analysis based on Poisson law. If a pipe that has been repaired has an individual failure rate higher than that of its group, its individual failure rate is retained in the results.

These tools, integrated in the CARE-W system, are available as software running under MS Windows® 2000 or later.

2.1.2 Description of the hydraulic reliability tools

CARE-W tools that allow the assessment of hydraulic criticality index (HCI) are:

- *Failnet-Reliab*, produced by Cemagref (France),
- *Aquarel*, produced by SINTEF (Norway),
- *Relnet*, produced by Brno University (Czech Republic).

They aim to assess pipe hydraulic importance. They are differentiated according to the type of models used and to the consideration of failure rate in the computation of HCI (Table 2.3).

Failnet-Reliab uses a specific hydraulic model. Contrary to classical models, which calculate the pressure at each node using user-defined water demands, this particular model is based on the calculation of available consumption, depending on pressure and predetermined water demand. Computation of HCI (Hydraulic Criticality Index) is based on the impact of pipe failure on this available consumption.

For instance, for the link j, the HCI is computed thus:

$$HCI_j = \sum_{i=0}^{n}(d_i - c_{ij}) \times t_{cj} \text{ (in m}^3\text{/year)}$$

with n, the number of consumption nodes on the network, d_i water demand at the node i, c_{ij} available consumption computed at the node i when link j fails, t_{cj} forecasted repair duration in the year.

This duration is assessed from forecasted failure rates (computed by *CARE-W_Poisson* or *CARE-W_PHM*) and assumptions on mean time to repair (MTTR). In this case, the computed HCI corresponds to non-supplied water volume in the year. It can be divided by total supplied water volume in the year to yield a value varying between 0 and 1. A HCI equal to zero means that the pipe has no effect on hydraulic reliability, either because its forecasted failure rate is equal to zero or because its unavailability (repair of failure for instance) will have no effect on customer interruptions to supply. A HCI equal to 1 means that the pipe is totally unavailable and that its unavailability will result in supply interruptions for all customers served. Even if this value cannot be reached, this 0-1 scale allows one to compare pipes, which belong to different hydraulically independent networks.

Aquarel and *Relnet* use the EPANET hydraulic model. The model is a classical model in which pressures are calculated according to demands imposed at each node by the user. As with *Failnet-Reliab*, *Aquarel* computes reliability indices, by integrating failure impacts of each pipe. The impact is, in this case, assessed according to the pressure drop at each node. The pipe availability is also considered, integrating failure rate and mean time to repair (MTTR). The final reliability index is a value between 0 and 1. The HCI for a pipe is the relative value of its hydraulic importance. This is done by renormalising the reliability index such that the pipes with the highest value of this index are given a HCI = 1. Other pipes receive an HCI, which is the ratio (pipe reliability index /_highest reliability index).

Another feature of *Aquarel* is that hydraulic modelling can be made over 24 hours and the water level in tanks taken into consideration.

Relnet does not consider failure rate and is based only on hydraulic computation. The final index is a value varying between 0 and 1. In this case an HCI equal to 1 means that a pipe break will result in interruptions to supply to all customers served by that pipe. *Aquarel* and *Failnet-Reliab* approaches are very similar (Figure 2.1).

2.2 SCIENTIFIC METHODOLOGY APPLIED

This task has several objectives:

- to test and compare the models on several water networks, that have different characteristics (size, geographical specificities, hydraulic conditions, material, type of data, maintenance data),
- to have a critical look on the models, with the aim of validating and fitting them,
- to improve their use and make it easier,
- to help to define a best use procedure.

2.2.1 FAIL tools

2.2.1.1 Objectives of the tests for FAIL tools

The objectives are to test whether the models fit the reality. This has been done either on whole networks or on categories of pipes (principally by material).

For each network, where historic maintenance data have been recorded for n years, the model was parameterised with the sole first n-x years and the results were compared to the x last years. (Cf. Figure 2.1)

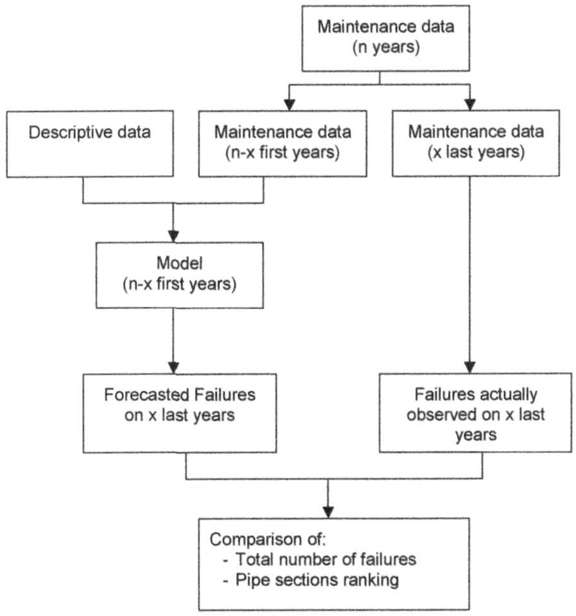

Figure 2.1. Comparison of forecasted vs. actually observed failures

For each model several tests were done with various combinations of:

- the length of the historic (n),
- the length of the time horizon (x).

For "Benefit" index (Cf. 0), one checked whether the results were better on the very first links (1% for instance, that is close to classical rehabilitation rate) or on a larger set (10, 25 or 50%).

A difficulty of this test is caused by the disappearance of the data related to replaced segments. This could make artificially depart the models from the reality.

The tests aimed to compare the results of different models. Different indices were used. This comparison was done on:

- rankings,
- influencing factors.

It was also done either on whole networks or pipes categories.

2.2.1.2 Estimated indices

Several different validations were performed.

For forecasting models, the validation consisted in:

- comparing the forecast results (with complete or uncompleted data) with the reality ; Two indices were proposed: one that gives a "Mark" to the model and one, more practical, that gives the benefit of the choice of the pipes to be replaced in term of reduction of the number of failures;
- comparing the results of 2 different models (or complete model to model with uncompleted or uncertain data); the comparisons of the two previous indices are used.

For each comparison model-reality and model-model, the indices values were calculated and these values were tested against the null hypothesis according to which the results are purely random.

"Mark" index

One interesting problem, which has been addressed neither in the statistical literature nor in the practical Cemagref studies, consists in building a tool to measure the usefulness of a given covariate, from the point of view of the forecasting ability of the model.

It is here proposed to build first an index to characterize the efficiency of the model to forecast the failure risk. In the sequel, this index is called the Failure Risk Forecasting Efficiency (FRFE) and denoted Φ, and can be computed in the model validation phase previously described. The model validation phase can then be performed twice:

a first time with the complete model, *i.e.* using all covariates found as being significant in the calibration phase, including the covariate which contribution to the forecasting ability of the model is to be assessed,

and a second time with the reduced model, *i.e.* deprived of the given covariate.

If the FRFE obtained with the complete and the reduced model are respectively denoted Φ_+ and Φ_-, the difference $\Phi_+ - \Phi_-$ measures the contribution of the given covariate to the forecasting ability of the model.

It is proposed to compute the FRFE as follows. Let first the random variable N_i stand for the number of failures the ith section may be subjected to in the time interval $[t_v, t_{ib}]$. The number of failures actually observed in this interval is denoted by $\widetilde{N}_i$, and the expected value (*i.e.* forecasted by the model) by $\hat{N}_i$. The c sections are then ranked in three ways:

the first ranking consists in sorting out the sections by descending values of expected numbers of failures $\hat{N}_i$ and the resulting ranks are denoted by $\hat{R}_i$; this means that:

$$\hat{N}_k = \underset{i=1...c}{Max}(\hat{N}_i) \Rightarrow \hat{R}_k = c \text{ and } \hat{N}_k = \underset{i=1...c}{Min}(\hat{N}_i) \Rightarrow \hat{R}_k = 1 \text{ ;}$$

the second ranking consists in sorting out the sections by descending values of observed numbers of failures $\widetilde{N}_i$ and the resulting ranks are denoted by $\widetilde{R}_i^+$; this means that:

$$\widetilde{N}_k = \underset{i=1...c}{Max}(\widetilde{N}_i) \Rightarrow \widetilde{R}_k^+ = c \text{ and } \widetilde{N}_k = \underset{i=1...c}{Min}(\widetilde{N}_i) \Rightarrow \widetilde{R}_k^+ = 1 \text{ .}$$

the third ranking consists in sorting out the sections by ascending values of observed numbers of failures $\widetilde{N}_i$ and the resulting ranks are denoted by $\widetilde{R}_i^-$; this means that:

$$\widetilde{N}_k = \underset{i=1...c}{Max}(\widetilde{N}_i) \Rightarrow \widetilde{R}_k^- = 1 \text{ and } \widetilde{N}_k = \underset{i=1...c}{Min}(\widetilde{N}_i) \Rightarrow \widetilde{R}_k^- = c \text{ .}$$

The quantity $\sum_{i=1}^{c} \hat{R}_i \widetilde{N}_i$ can take any integral value between $\sum_{i=1}^{c} \widetilde{R}_k^+ \widetilde{N}_i$ and $\sum_{i=1}^{c} \widetilde{R}_k^- \widetilde{N}_i$.

It is then proposed to define the FRFE as:

$$\Phi = \frac{\sum_{i=1}^{c} \hat{R}_i \widetilde{N}_i - \sum_{i=1}^{c} \widetilde{R}_k^- \widetilde{N}_i}{\sum_{i=1}^{c} \widetilde{R}_k^+ \widetilde{N}_i - \sum_{i=1}^{c} \widetilde{R}_k^- \widetilde{N}_i} \quad (7)$$

The FRFE index Φ has the property: $\Phi \in [0,1]$. If the model produced a perfect forecast, the ranks $\hat{R}_i$ and $\widetilde{R}_i^+$ should be equal for all sections, and thus $\Phi = 1$, which means that the perfect model has a forecasting efficiency of 100 %. It is important to notice that FRFE does not measure the exactness of the forecasted numbers of failures, but rather the ability to correctly rank the sections according to their actual risk of failure.

It remains to carry out the theoretical investigation of the distribution of the FRFE considered as a random variable Φ_0 under the null hypothesis H_0 of independence between $\hat{R}_i$ and $\widetilde{R}_i^+$. This would then make it possible to compute the risk $P\{\Phi_0 > \Phi \mid H_0\}$ to reject wrongly the null hypothesis when asserting the forecasting efficiency of the model.

Benefit index

This index aims at assessing the number of failures avoided, if the links with the highest break probabilities were rehabilitated.

Let first the random variable N_i stand for the number of failures the *i*th section may be subjected to in the time interval $[t_v, t_{ib}]$. The number of failures actually observed in this interval is denoted by $\widetilde{N}_i$, and the expected value (*i.e.* forecasted by the model) by $\hat{N}_i$. Contrarily to the previous index, the c sections are then ranked as follows:

- the sections are sorted out by ascendant values of expected numbers of failures $\hat{N}_i$ and the resulting ranks are denoted by $\hat{S}_i$; this means that:

$$\hat{N}_k = \underset{i=1...c}{Max}(\hat{N}_i) \Rightarrow \hat{S}_k = 1 \text{ and } \hat{N}_k = \underset{i=1...c}{Min}(\hat{N}_i) \Rightarrow \hat{S}_k = c \text{ ;}$$

For each pipe we can then calculate the rank percentage $\hat{p}_i$: $\hat{p}_i = \frac{\hat{S}_i}{c}$

Let $\widetilde{F_i}$ be the number of failures actually observed in the i segments that have the highest failure probability:

$$\widetilde{F_i} = \sum_{k/\hat{S}_k<i} \widetilde{N_k}$$

And $\widetilde{f_i}$ the percentage of failures actually observed in the i segments that have the highest failure probability:

$$\widetilde{f_i} = \frac{\widetilde{F_i}}{\sum_{i=1}^{c} \widetilde{N_i}}$$

It is then possible to draw the graph (see example) of $\widetilde{f_i}$ according to $\hat{p}_i$ and we propose to assess different indices $\widetilde{f_i}$ according to $\hat{p}_i$ for instance 1%, 5%, 10%, 25% and 50%. The higher these values are, the better the model is.

2.2.1.3 Global synthesis for each test

Integrating the 2 models and a 'simple' model (with simple ranking and assessment) and presenting the indices assessed for each test and each model.

One test depending on:

- the model,
- the failure observation time (which can be increased or decreased),
- the considered factors (environmental factors essentially).

Table 2.3. Synthesis of the different tests of FAIL tools

	Results compared	**Used indices**
Test model vs. reality	Ranking Total number	- Mark Index - Benefit Index
Test model vs. model	Ranking Total number of failures Influencing factors	- Difference of Mark indices
Test uncompleted data model vs. complete model according to reality	Ranking Total number of failures	- Mark Index - Benefit Index - Difference of Mark indices

2.2.2 REL tools

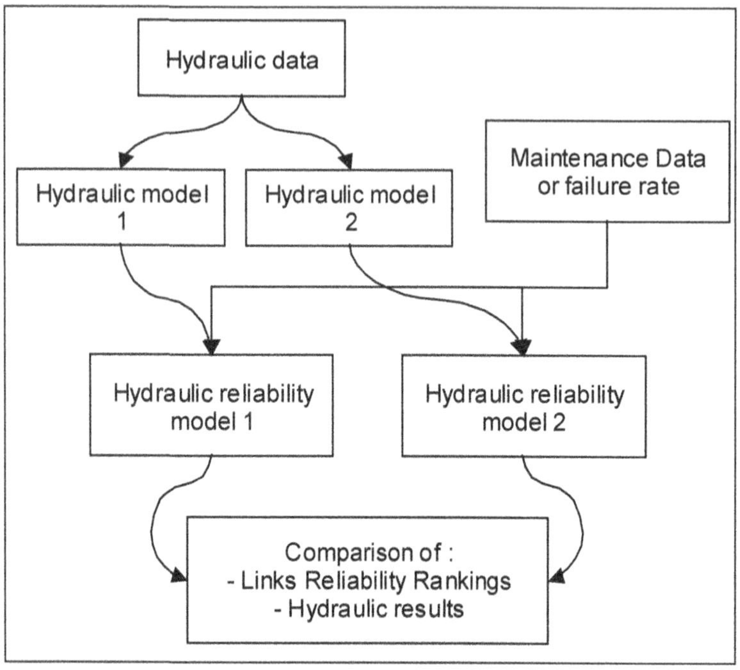

Figure 2.2. Comparison of Hydraulic Reliability Models

The models have been compared according to different indices (cf. Figure 2.2). The comparison has been done on:

- hydraulic model results

This comparison was made on important consumption nodes of each network. It aims to eventually detect causes of divergence in reliability results.

- hydraulic reliability rankings of the links.

Basic hydraulic parameters (duration, demand, desired pressure, roughness) were the same for each model.

2.3 SCIENTIFIC RESULTS OBTAINED

All the tests and results are presented in the report D4 (Eisenbeis et al, 2003)

2.3.1 Tests on FAIL tools

The tests on FAIL models have been made on 3 networks:

- Trondheim in Norway (Poisson, PHM, Winroc),
- Stuttgart in Germany (Poisson, PHM),
- Lausanne in Switzerland (PHM).

2.3.1.1 The data

The specifications of the water services are presented in Table 2.4.

Table 2.4. Characteristics of the networks for FAIL tools tests

	Trondheim	Stuttgart	Lausanne
Length (km)	827	1424	734
Number of pipe segments	7659	19195	8126
Material	Ductile Iron (50%) Cast Iron (28 %) PVC (8 %) Steel (4%) Abestos cement (4%) +Concrete, Copper, PE	Ductile Iron (48%) Cast Iron (45 %) PE (4.8%) Steel (2%) PVC (0.2 %)	Ductile Iron (58%) Cast Iron (22%) Steel (12 %) Iron (3%) Asbestos cement (3 %) PE-PVC (2%)
Diameter	19 – 4000 mm	50-1000 mm	19-700 mm
Variables	Type of soil	Internal and external protection, type of joint, observed depth, bed, corrosion, deposit, type of failure	Water pressure, Type of supply,
Failure observation time	1988-2000	1985-2000	1980-2000
Number of failures	2304	2900	980

Installation date

Trondheim and Lausanne Networks have the same installation date profiles:

- 100 km on the total laid up to the second world war,
- a development of the networks from 1945-1950,
- a higher development in the 1960's and 1970's.

In Stuttgart, a greater part of the network was laid in 1940 (almost 30 % of the network). Just after the war, a constant length of pipe has been laid up to now.

Material

About Ductile Iron and grey cast iron, Trondheim and Lausanne have again the same description. Half of the pipes are in Ductile Iron and around 20 to 28 % are in Grey Cast Iron. For Stuttgart the grey cast iron pipes take a greater part: 45 %. This characteristic is linked to the installation date.

A characteristic of Lausanne is its high number of steel pipes (10 %).

Diameter

In the three networks, the majority diameter is 100 to 150 mm. In Stuttgart, the part of diameter between 175 and 250 is greater, due to the size of the city.

Failures by year

In Trondheim and Stuttgart, the number of failures is relatively constant for 15 years. In Lausanne, this number was constant up to 1995 and has been increasing for 5 years. This increasing is mainly linked with a better recording of maintenance data as failures and the setting up of a computer database.

2.3.1.2 Results

Table 2.5 synthesises the result of Mark Index for several forecasts, different according to the water services and the length of record periods. Several comments can be done:

- For the three services, the Mark index is around 80%, which shows a good efficiency of the models (excepted for PHM model, with Stuttgart data).
- The length of record period has not always the same action. In Trondheim shorter the period is lower is the Mark Index. In Lausanne for DGCI as for steel pipes, shorter is the period, higher the Mark Index is. In Stuttgart, the results are contrary according to the chosen model. In conclusion, the results show that a

longer period is better, but it is indispensable that these records are reliable. For instance in Lausanne the records exist since 1980 but it has been viewed that only records since 1995 are really reliable.

- PHM and Poisson give similar results concerning Trondheim. However results were less efficient in Stuttgart than in other networks with PHM model. The reason for these results was not very obvious. It has been assumed that it was possibly due to a rehabilitation policy existing in the service for now a long period (20 years) with strict rules to replace pipes, that means that pipes in bad condition are still existing.

Table 2.5. Results of Mark Index according to the water services and length of failure record period (DGCI means Ductile and Grey Cast Iron)

	Failure record Period	PHM	Poisson
Trondheim	1988-1995 (DGCI)	0.826	0.833
	1990-1995 (DGCI)	0.818	0.811
	1992-1995 (DGCI)	0.811	0.814
	1994-1995 (DGCI)	0.793	
Stuttgart	1985-1995 (DGCI)	0.717	0.819
	1990-1995 (DGCI)	0.675	0.812
	1994-1995 (DGCI)	0.774	0.809
Lausanne	1980-1998 (DGCI)	0.824	
	1995-2000 (DGCI)	0.838	
	1980-1998 (Steel)	0.788	
	1995-2000 (Steel)	0.862	

The analysis of Benefit indices confirms these results. Figure 2.3 shows the benefit indices for the different period with PHM modelling.

These results show that the forecast is less efficient with a very short period (less than 2 years). However, a 2 years observation period still provides a relatively good forecast. Indeed 5% of the pipes with the risk of highest failure can avoid almost 25% failures and 10% allow to avoid almost 40%.

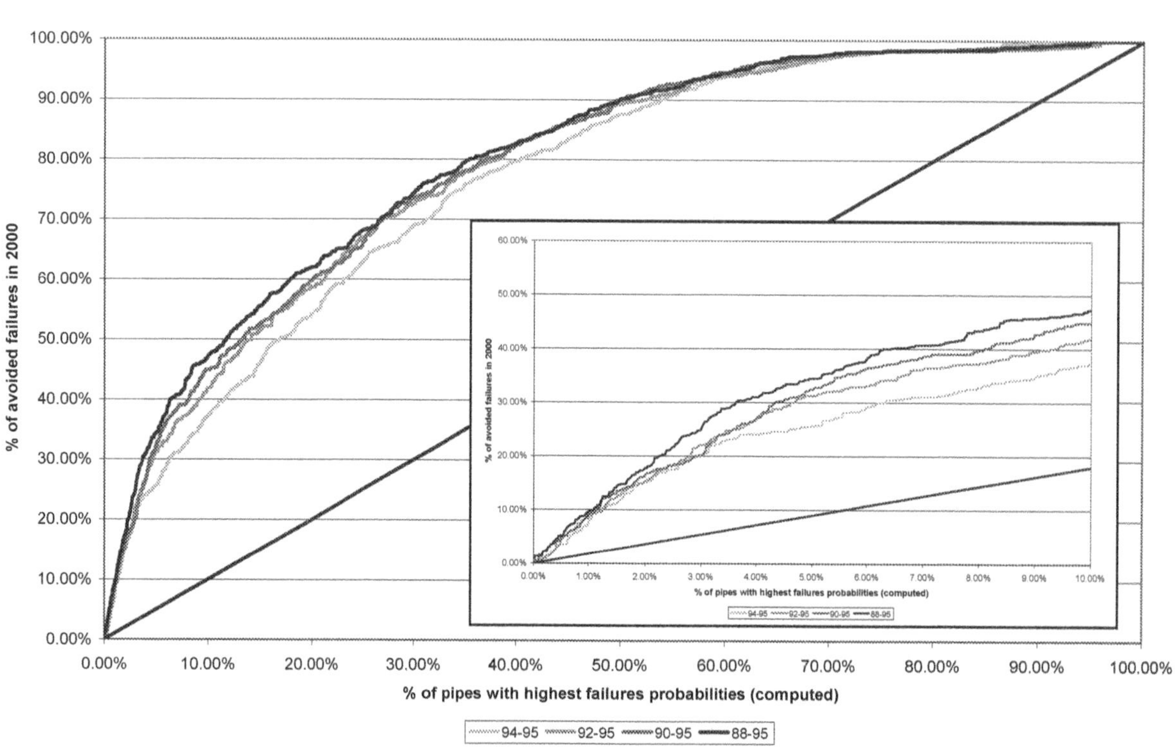

Figure 2.3. Comparison of Benefit Indices according to data period used to elaborate the model (DGCI) (Trondheim)

- Influence of the variable "soil" on the forecast.

The environmental variable "soil" was available in Trondheim network. These tests aimed to assess the influence of this variable. For each model, two tests have been applied, with and without the variable "soil".

The mark index provides a little improvement by considering the variable "soil" for PHM as for Poisson. However the benefit index doesn't show an evident improvement especially for the pipes with highest failures risk. Variable soil contributes mainly to distinguish the pipe with low failure risk, i.e. without previous failures.

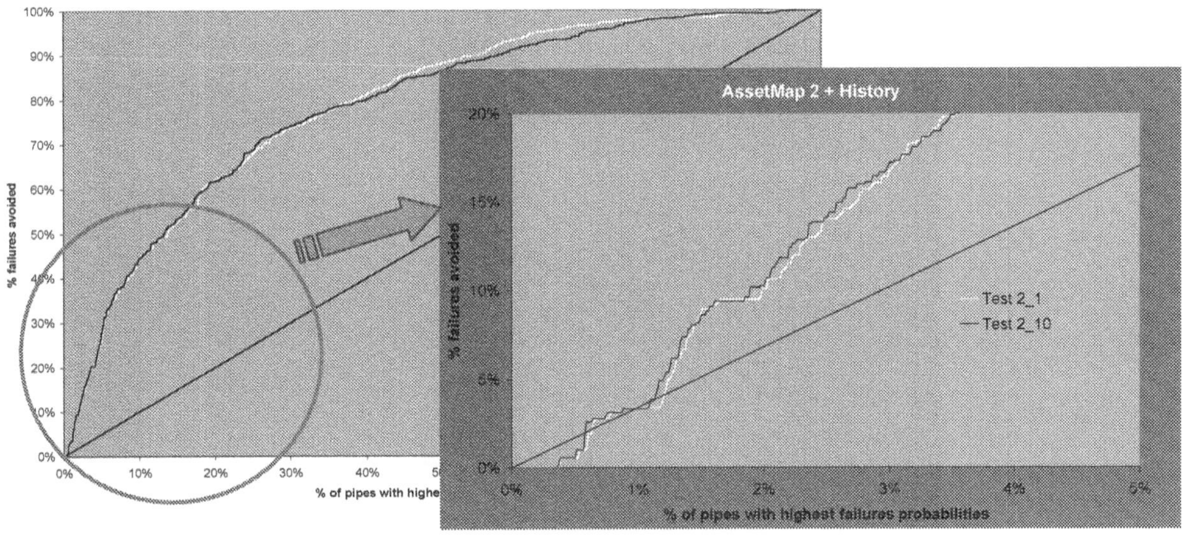

Figure 2.4. Comparison of Benefit curves, using or not soil variable (Poisson, model 1988-1995 compared to 1996-2000) (Trondheim)

Table 2.6 presents a synthesis of the significant variables of the Poisson model according to the type of failures. It gives several indications:

- The failures due to corrosion appear mainly for the little diameter pipes and aged between 20 and 30 years. Indeed the pipes more than 30 have less corrosion than 20 to 30 years. This shows this more linked to a specific laid period than to the ageing of the pipes.
- The circular breaks appear mainly on little diameter pipes in Grey Cast Iron.
- On the other hand, longitudinal breaks appear on large pipes in Ductile iron.
- Globally failures on pipe appear on aged little pipes in Grey Cast Iron and failures on joint are mainly linked to the age of the pipes.
- Tests according to the period length of data record

Table 2.6. Significant variables according to the type of failure (Poisson Model, Stuttgart)

Material	DI-CI	DI-CI	DI-CI	DI-CI	DI-CI	Steel	PE – PVC
Type of failures	**Corrosion**	**Circular**	**Longitudinal**	**Pipe failures**	**Joint failures**	**All failures**	**All failures**
Diameter							
<= 80		1		1			
] 80-100]	1		1		/	1	/
] 100- 150]		0.484		0.764			
> 150	0.481	0.224	1.856	0.362		0.359	
Material							
CI	/	1	1	1	1		
DI		0.112	31.075	0.343	/		
HDPE PE PVC							1
MDPE							0.252
Age							
[0-10]	1	1		1			1
[10-20]	3.169			2.532	1		
[20-30]	10.644	10.113	/	9.001		/	2.600
[30-40]	6.159	9.052		7.405			2.488
> 40		9.810		9.278	3.461		

Tests made in the Care-W project *(Eisenbeis et al, 2003)* have shown that a simple classification of the pipes according to their past failure numbers could also give interesting results to classify pipe to prioritise. (Cf. Figure 2.5). However it has been noted that only 10 to 15 % of the pipes have broken in the past. Consequently with this kind of classification all the pipes without previous breaks (at least 80 to 85 %) cannot be classified. The same figure shows that without a model (yellow curve) it is not possible to classify pipes beyond 10 to 15%, contrary to Poisson and PHM Model. This is especially so for services that have short recording period times (Cf. Figure 2.5).

Without Poisson and PHM models, it is then difficult to use the Failure Rate as parameter for Care-W_ARP. An example of a criterion used in the multi-criterion decision support software (CARE-W_ARP) shows that the assessment of the failure rate is important not only for high values but also for low values.

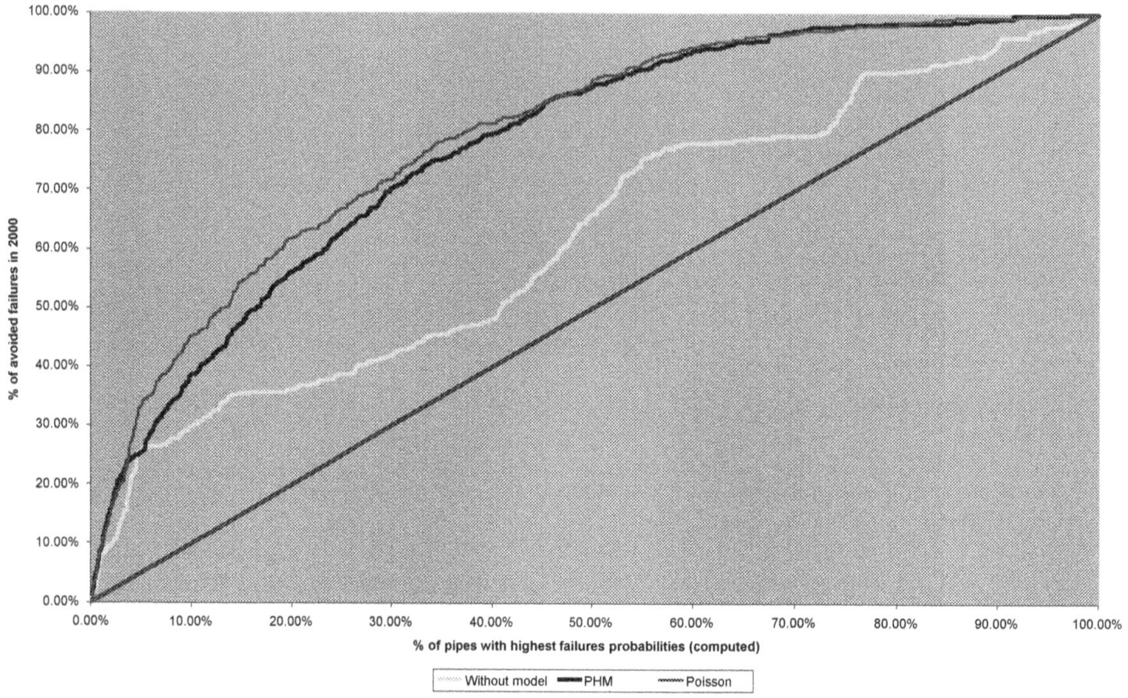

Figure 2.5. Benefit index of Care-W PHM, Care-W_Poisson and without model (short record time, Trondheim)

As the consequences of a failure may be much higher for some particular pipes, rehabilitation of pipes with low failure rates may appear as priority projects within a preventive approach.

Criterion PWI (Predicted Water Interruption) is calculated as follows:

PWI = PBR. EDI. NPS

With:

PBR: Predicted burst rate

EDI: Expected duration of interruption (hours)

NPS: Number of people supplied by the link

Figure 2.6 (Reggio Emilia case study) shows that a pipe with a low burst rate (e.g. 0.02 bursts/100m/year) may appear as a highly ranked rehabilitation candidate, similar to other pipes with a high burst rate (e.g. 0.5 bursts/100m/year).

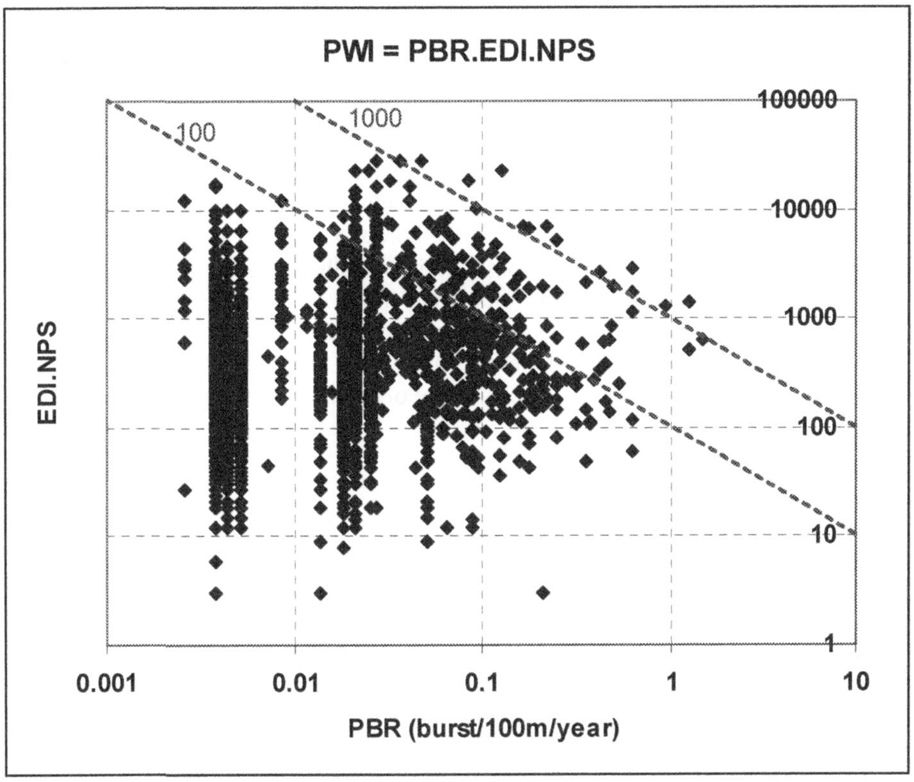

Figure 2.6. Interest of considering pipe with low predicted burst rate

2.3.1.3 Conclusion on FAIL tool tests

The tests made on the different networks have shown that the FAIL tools (Poisson and PHM) can bring to the water supply benefits concerning the choices of the pipes in the frame of annual rehabilitation program. Notably it has been shown that in some of the cases, choosing 10% of the pipes declared as with the highest failure risks could allow to avoid 50% of futures failures. These benefit indices could be fitted considering the criteria defined in ARP module and that used failure rate or failure forecast in their calculation (Repair costs, disturbances, etc.)

Secondly the influence of the existing data has been assessed. The variation of the data maintenance period used to build the model shown several results:

- long maintenance data period doesn't give always suitable results.

This is due first to the disappearance and the replacement of pipes, which were in bad condition and included in rehabilitation programs. Consequently short maintenance data will show a better image of actual pipe condition. Last, because of the setting up of GIS or computer database, the data are better recorded and in a more rigorous way.

- the consideration of environmental variable:

Only one test has been done. This concerned the soil considered on Trondheim networks. This didn't show a real improvement of the forecast even if the mark index was slightly higher with considering the variables. However, this type of variable could in other cases contribute to a better understanding of the actual pipe deterioration and thereby the expected service life.

Third concerning the significant variables, the tests confirmed the results of previous studies: the length, the material, the installation period, the diameter and the number of previous failures has almost always significant, with a relative risk close to values assessed previously.

Fourthly in one case (Stuttgart with PHM model), results were less efficient than in other networks. The reason of these results was not very obvious. It has been assumed that it was possibly due to a rehabilitation policy existing in the service for now a long period (20 years) with strict rules to replace pipes, that means that pipes in bad condition are still existing.

Fifthly a comparison made with simple classification of the pipes according number of previous failures and diameter has shown that this kind of "simple model" could give also interesting results. However this type of approach doesn't allow an efficient calculation of future failures useful to assess the criteria of ARP, notably because of an uncertainty higher than Poisson and PHM. Also this approach doesn't allow differentiating between the pipes on long-term approachs.

2.3.2 Tests on REL tools

2.3.2.1 Ugla, Trondheim

The Ugla zone is a small zone in Trondheim. The zone consists of 201 pipes, a tank and a pumping station. In order to make computable the zone hydraulic, a reservoir has been included in the model downstream the pumping station (supplying the pumps). In reality there is no reservoir located here. At the border of the zone there are several closed pipes. These pipes are closed under normal operations. These pipes lead into another pressure zone. If needed (e.g. in case of pipe failure) these pipes can be opened manually by the operating crew.

For the reliability simulations it is assumed that the pressure should be more than 25 meters (100% pressure). The reliability analysis is also carried out for the peak demand of water consumption. The maximum demand factor is 1.3. All the analysis starts with this flow as the initial condition (i.e. in case of extended time simulations).

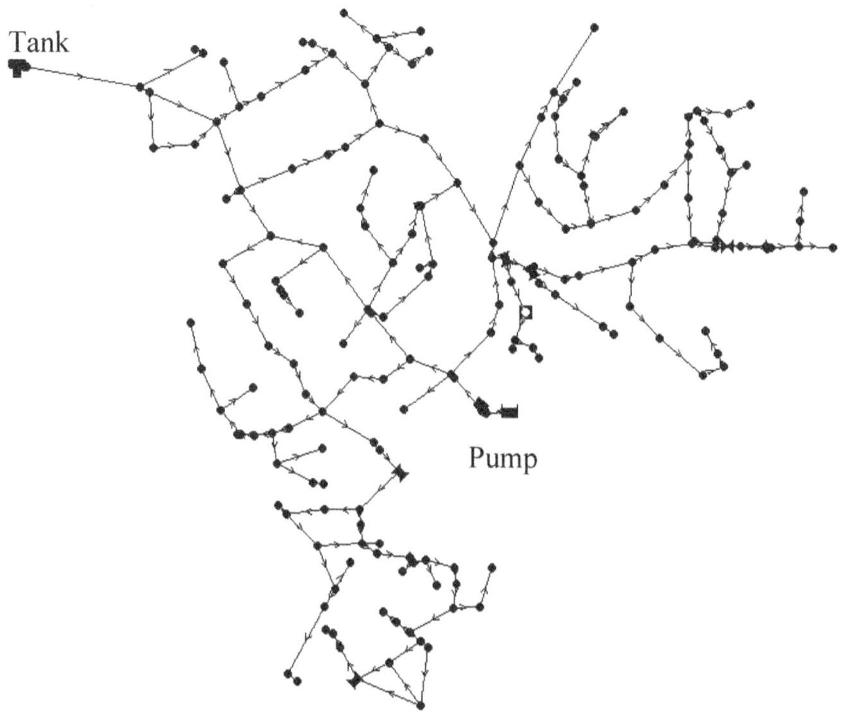

Figure 2.7. Illustration of the water zone Ugla in Trondheim

To evaluate the unavailability time of a pipe, linked to a failure, the assumption presented in Table 2.7 has been made. This signifies that the repair time of a pipe depends on the diameter. This assumption could be sharpened according to other factors like, for instance, the pressure or the traffic in the street and according to the knowledge of water supply personnel.

Table 2.7. Repair times for pipes in Trondheim

Pipe diameter (mm)	Repair times (hours)	
	Mean Time To Repair (MTTR)	1% TTR
< 300	8	24
300-400	16	48

Aquarel

For Aquarel three simulations have been made:

1. Aquarel with the effect of tanks but with constant water consumption in each node (default Epanet)
2. Aquarel with the effect of tanks and with sprinklers ([emitters]) in each node allowing for pressure dependent water consumption (to be compared to Failnet-Reliab)

3. Aquarel without the effect of tanks but with constant water consumption in each node (old Aquarel version) (in order to compare with the other models)

In these three simulations it was possible to classify the consumption nodes according to the availability of water. The Figure 2.8 shows the 20 weakest nodes according to this availability. These nodes are :

- nodes located closed to tanks,
- nodes located on branches.

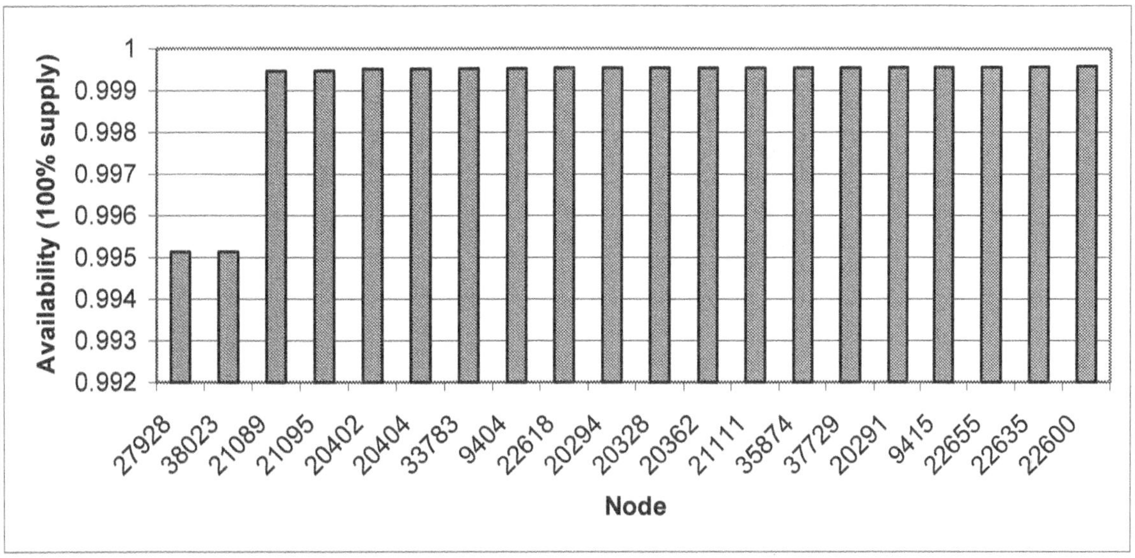

Figure 2.8. Availability of the 20 "weakest" nodes

Concerning the importance of the pipes, the results are different according to the tests.

For the test with effect of tanks and constant node consumption, the pipes considered as the most important are pipes located near the tank (Cf. Figure 2.9).

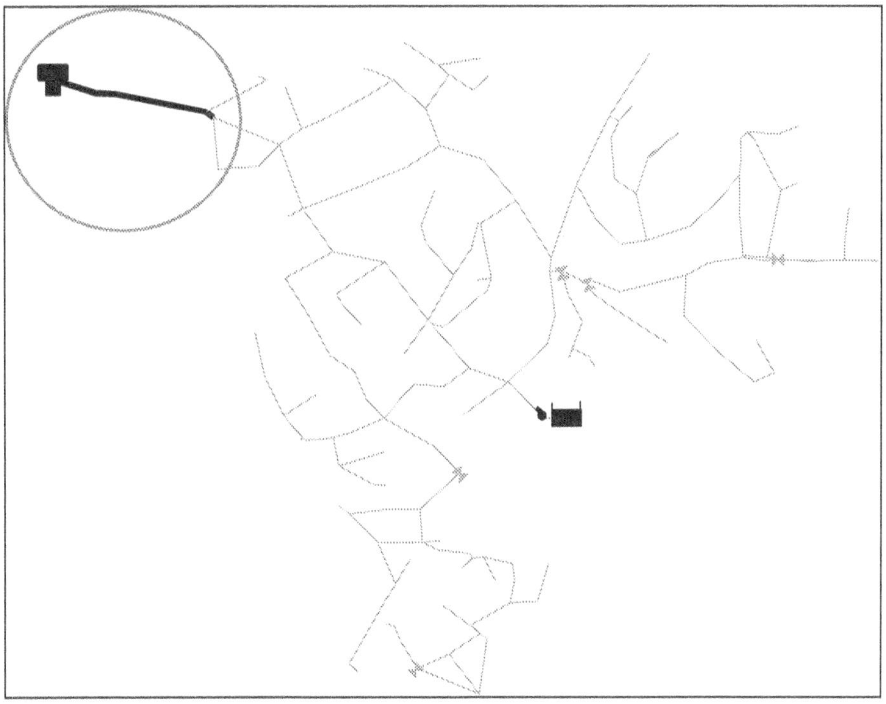

Figure 2.9. Important pipe for Aquarel model (with effect of tanks, constant consumption)

For the tests with effect of tanks and with sprinklers (simulating pressure dependent consumptions), the important pipes are different. These are mainly the pipes at the beginning of branches.

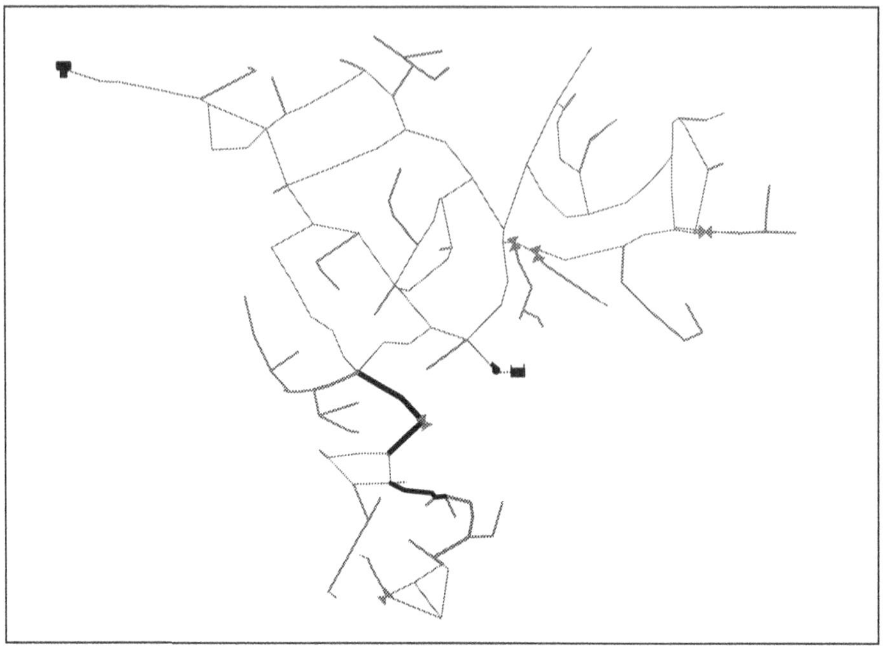

Figure 2.10. Important pipes for Aquarel model (with effect of tanks, sprinklers)

To conclude, the different important measures are calculated and ranked with respect to each measure. The unavailability measure and the frequency importance measure reports for this network the same relative ranking. The Aquarel version without sprinklers reports more or less the same ranking of the pipes. The most important pipes are the pipes downstream from the tank. When sprinklers are included in the simulations a more sophisticated picture appears. Important pipes are than also discovered in the southern part of the network. These pipes are typically the mains into a smaller closed loop in the network.

Failnet-Reliab

Two types of results, useful for Annual Rehabilitation program, are proposed with Failnet-Reliab. These are:

- results concerning pipe, giving an hydraulic criticality index, considering at the same time, the effect of the failure of the pipe and its failure risk,
- a global reliability index, defining the global hydraulic vulnerability of the network linked to failures.

Moreover it is possible to include failure risk forecasted from FAIL tools.

On Ugla Networks, results were close to Aquarel results with sprinklers (Cf. Figure 2.11). Vulnerabilities of the nodes have also been assessed and are presented on this Figure. It shows that the most vulnerable nodes are:

- nodes located on the branches,
- nodes located near to the tank, because of their elevation close to tank water level.

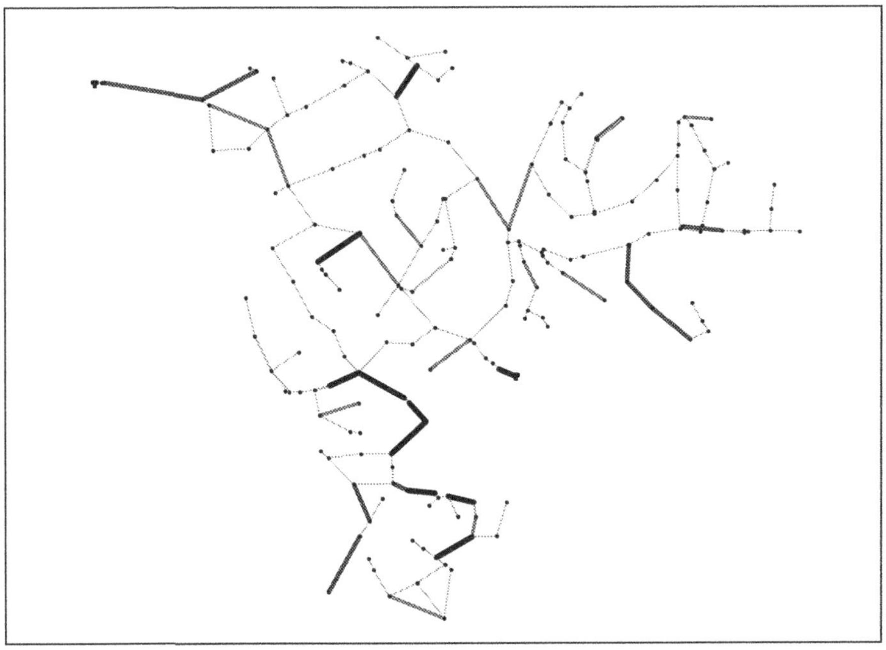

Figure 2.11. Failnet-Reliab results on Ugla zone (thickness and darkness of the pipe defines its importance)

The results of Figure 2.11 have been computed with a constant failure rate for all the pipes. If forecasted failure rates for each pipe are considered in the computation, the results are sensitively different (Cf. Figure 2.15). It shows at the same time pipes which have a high hydraulic importance and which have high failure risks.

However it must be noted that this network is very reliable. Indeed the maximum HCI index is equal to $1.02\ 10^{-5}$ for the first case and $2\ 10^{-4}$ for the second case. This is confirmed by the values of global reliability index that is equal to 0.974. This shows a network with a high number of loops, which is moreover supplied by 2 tanks.

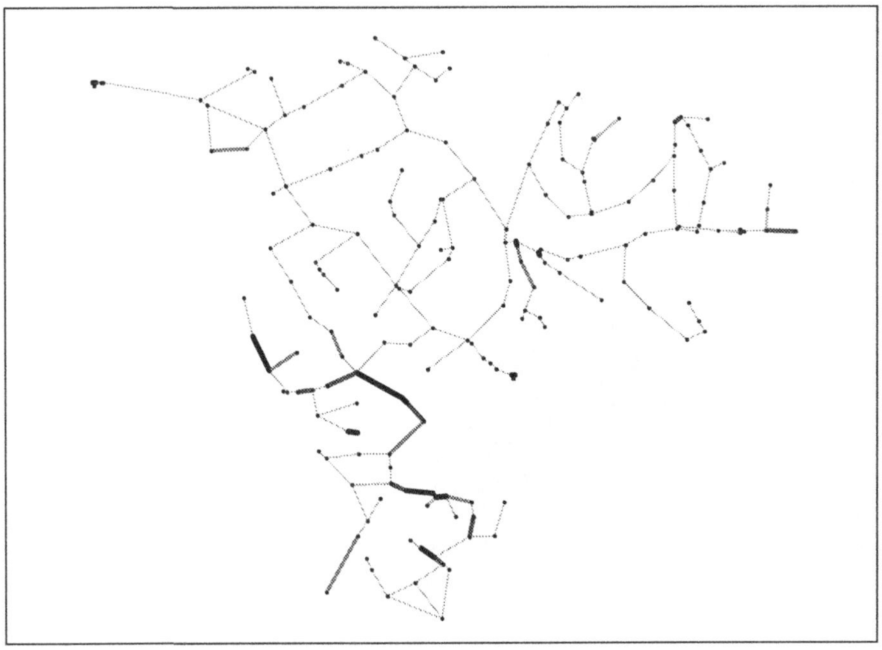

Figure 2.12. Failnet-Reliab results on Ugla zone considering specific failure rates by pipe (thickness and darkness of the pipe defines its importance)

Relnet

Relnet results on Ugla zone are presented in the Table 2.8. It shows that, as for Aquarel with constant consumption, the most important pipes are pipes located near to the tanks, and at the lower level, the pipes located at the beginning of the branches.

Table 2.8. The 20 worst pipes according to their hydraulic importance (Relnet)

Pipe section ID	Ph14b	Pipe section ID	Ph14b
12899	1.00000000	181447	0.10989011
184539	1.00000000	181450	0.10439560
191162	1.00000000	186040	0.09340659
169942	0.98351648	167474	0.08241758
167703	0.18681319	167486	0.07692308
167690	0.18131868	167488	0.07142857
187468	0.18131868	185065	0.07142857
167659	0.17582418	167701	0.06593407
181459	0.11538462	181471	0.05494505
181447	0.10989011	167699	0.03846154
181450	0.10439560	181481	0.03846154

The global reliability index equal to 0.99 confirms the very high level of reliability of the network.

2.3.2.2 *Crissier, Lausanne*

Crissier is a water zone of Lausanne water supply. It is made up of:

- 1130 pipes,
- 996 nodes,
- 1 tank.

Two parts of the network are distinguished (Figure 2.13):

- a urban part, more dense and looped,
- a semi-urban, less dense.

For each test, two computation assumptions have been made:

- for the pressure: desired pressure for each node is equal to 25 m (or 2.5 10^5 Pa),
- for the demand: for model made on one time step, the water demand is equal to the hourly maximal demand.

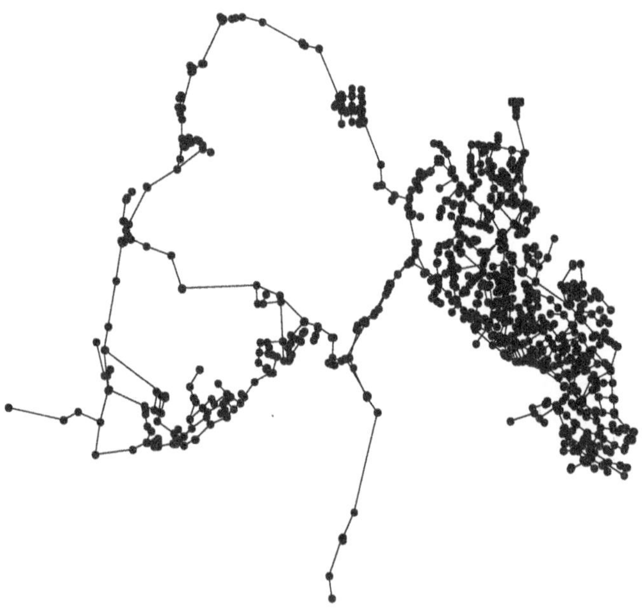

Figure 2.13. Illustration of the water zone, Crissier in Lausanne

Failnet-Reliab

Results coming from Failnet-Reliab about Crissier network are presented in the Figure 2.14, Figure 2.15 and Figure 2.16.

The Figure 2.14 and Figure 2.15 present the computation of Hydraulic Criticality indices. They show that most hydraulically important pipes are:

- the pipes located near the tank,
- the pipes located on branches,
- the pipes located at the beginning of the main loop.

Figure 2.14. Failnet-Reliab results on Crissier zone, same failure rate for all the pipes (thickness and darkness of the pipe define its importance)

Figure 2.15. Failnet-Reliab results on Crissier zone considering specific failure rates by pipe (thickness and darkness of the pipe defines its importance)

The comparison between the two figures doesn't show a big difference concerning the hydraulic importance, in spite of considering specific failure probability for each pipe. However some pipes appear as important, regarding the HCI, when specific failures risk is considered.

Figure 2.16 presents the more vulnerable nodes according to the calculation performed with Failnet-Reliab.

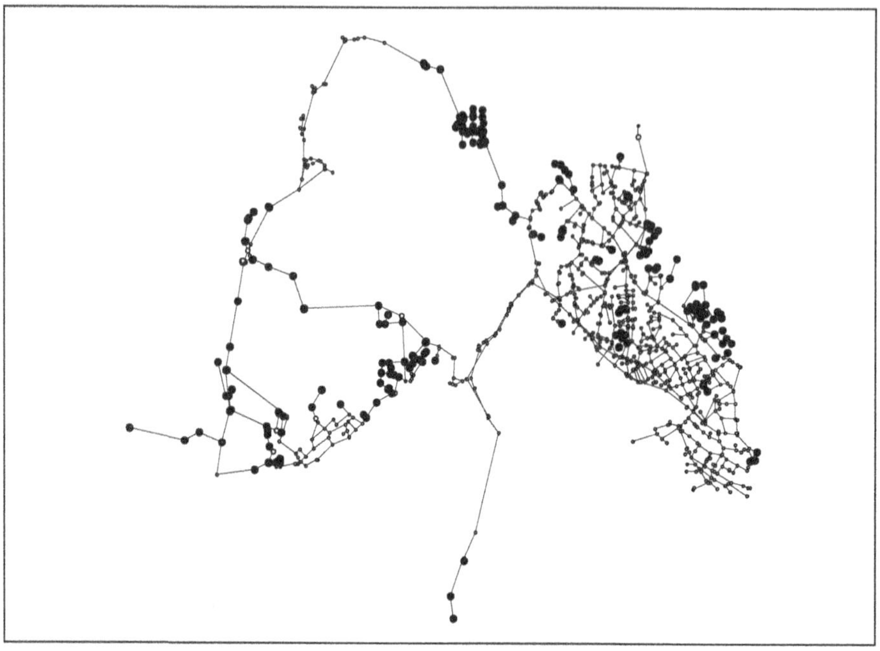

Figure 2.16. Failnet-Reliab results on Crissier zone considering specific failure rates by pipe (thickness and darkness of the nodes defines their vulnerability to the failures)

Relnet

For Crissier the results provided by Relnet are the number of nodes with a pressure below than the required pressure (25 m) and the index giving the impact of the failure of a pipe on the pressure in the node.

The Table 2.9 gives the 20 worst pipes in the sense of Relnet index.

If we compare with Failnet-Reliab results, we can see 8 pipes are considered commonly in the 20 worst pipes. The main difference is due to considering the length of the pipes to calculate their unavailability.

2.3.3 Conclusion

The tests have shown several approaches concerning the calculation of Hydraulic Criticality indices.

Aquarel allows the consideration of tanks and their emptying to calculate the indices. Moreover it allows to calculate indices according to 2 ways:

- with constant consumption at each node, in this case the indices will be based on node pressure,
- with "sprinkler": in this case the indices are based on the ratio between consumption and water demand.

Table 2.9. Classification of the 20 worst pipes according to Relnet and Failnet-Reliab

Relnet Classification				***F-Reliab classification***
LinkID	**Removed Nodes**	**Nodes<RP**	**Ph14b**	
632	994	0	0	*972*
972	993	1	0	*631*
1075	0	994	0	*632*
631	0	993	0.001	*897*
1063	0	992	0.002	*407*
897	0	991	0.003	*529*
896	0	990	0.004	*432*
1035	0	247	0.7515	*147*
669	0	246	0.7525	*669*
670	0	241	0.7575	*670*
673	0	239	0.7596	*777*
672	0	235	0.7636	*673*
914	0	163	0.836	*672*
1086	0	163	0.836	*776*
630	0	161	0.838	*738*
636	0	155	0.8441	*914*
913	0	152	0.8471	*381*
768	0	150	0.8491	*305*
912	0	144	0.8551	*774*
911	0	140	0.8592	*812*

Failnet-Reliab, which computes a consumption dependent of the head, gives results close to Aquarel with "sprinkler". Moreover considering at the same time the ratio between available consumption and demand and the failure forecast coming from FAIL tools, it allows to highlight pipes and to separate them in a specific way.

Relnet allows a simple calculation of indices based on the increasing of pressure, without considering the failure risks. However it provides also results close to Aquarel with constant consumption.

One major interest of these tools is to define the hydraulic importance of each link. This interest concerns "branch" networks, where this importance seems obvious as well as for “looped” networks. This interest is higher when the failure rate can be included in the computation.

Figure 2.17 shows that even for a "branch" network, the REL tools can bring useful information. The first figure classifies the pipe according to the diameter, the second one according to the HCI (without considering the failure rate) and the third one considering the failure rate.

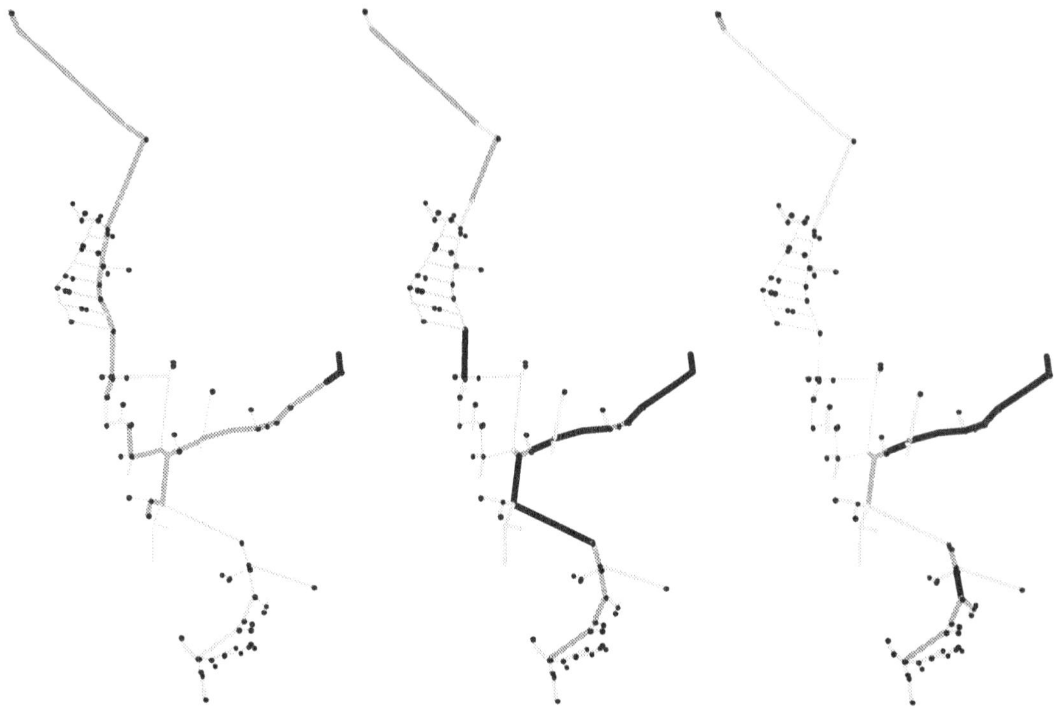

Figure 2.17. Comparison of results of REL tools (on the left: the thickness of the line represents the diameter, middle: HCl-Failnet-Reliab, without considering Predicted Failure Rate), right: HCl-Failnet-Reliab, considering Predicted Failure Rate)

2.4 PRACTICAL APPLICATION FORESEEN

Practical application foreseen for these technical tools is immediate: the tools are included in the CARE-W tools kit. It is closely linked to the CARE-W ARP (Annual Rehabilitation Program, see chapter 3).

With respect to failure forecast models, several scenarios may occur. If the water service provider already has a database with pipe attribute information including failure data, it is only a matter of adapting it to match the proposed format. This has been designed such that it is easy to use.

It is, however, useful to consider the following aspects:

- some data attributes relating to the pipes are indispensable, namely diameter and material;
- installation date is an essential parameter for burst assessment. Indeed, besides natural ageing, previous studies have shown that pipes laid in some periods, such as during the post-war period 1945-1950, were more likely to fail. This is also the case of the first ductile iron pipes,
- some environmental factors also have an influence on failure occurrences. This is the case for soil type, road traffic type or location of the pipe in the street (under pavement, – under road etc.). The use of Geographical Information Systems (GIS) can facilitate the consideration of this information.
- other parameters specific to water utilities can also be considered, particularly where correlations between these factors and failure data have been observed.
- with respect to failures records, the longer the history of data, the better the analysis of any trend in failures should be. However, previous studies[(5)] show that, in some cases, recent data are preferable to longer historical records. Recent data may be more consistently recorded and reliably collected. There may have been a change in reporting procedures and practices or losses from data transfer to newer replacement data management systems, which means that direct comparisons with older datasets are impossible.

If a service is to set up network maintenance databases in a systematic way, it is advisable to use record forms which can be used to describe accurately the failure date, type and cause, and provide information on pipes and their environment during the repair (e.g. material, diameter, soil type, lining condition, bedding materials)[(7)].

These data may then be used by *CARE-W_PHM* and *CARE-W_Poisson*. Tests performed in the CARE-W project have shown in a short space of time that, depending on water utility size, a minimum of 3 to 5 years of records should allow a reasonable or good failure forecast. This was deduced from a comparison of forecasted and observed failures[(5)].

Table 2.10 summarises the different aspects of CARE-W PHM and CARE-W Poisson and thus helps the user to choose the most appropriate.

Table 2.10. Characteristics of Care-W_PHM and Care-W_Poisson

	Poisson	PHM
Pipe Data	Indispensable: diameter and material, installation date (approx.) Useful: other environmental pipe data	At least, diameter, material, installation date (approx.) Useful: other environmental pipe data
Failure Data	2-3 years	2-3 years
Objectives:		
- Variable significance test	Yes, with statistical software	Yes
- Failure rate by category	Yes	Not directly
- Failure rate by pipe	Yes	Yes
- Forecast	No	Yes

The choice of the REL tools will depend on several aspects (cf. Table 2.11):

- **the existing model**: if a hydraulic model exists, it will be advisable to use Aquarel and Relnet if EPANET was used and to use Failnet-Reliab if Porteau was used. In other cases, no advice is proposed as the two file formats are Text-files and can be easily created,
- **the knowledge of pipe failure rate or predicted pipe failure rate** (for instance from PHM or Poisson): in this case it will be advised to use Aquarel and Failnet-Reliab that can include these values in HCI computation. If pipe failure rate is not known, category or general failure rate can be used, **the possibility of calculation on 24 hours (Aquarel),**
- **calculation time:**

Calculation time is different according to the tools. Relnet and Aquarel based on Epanet, are really more rapid. Failnet is slow. This is due to:

- the calculation model (head dependent model), which is a little more complicated than a classical model,
- the computer program, which has not been optimised. Several improvements are possible. The program has indeed been developed in 'Matlab' language and automatically translated in C++ language.

Table 2.11: REL tools characteristics

Feature	Aquarel	Failnet-Reliab	Relnet
Hydraulic model	Classical model (EPANET) on 24h, considering tank water levels	Head dependent demand model (based on "Porteau" Model)	Classical model (EPANET)
Equipment integration	Yes	In progress	Yes
Failure rate integration	Yes	Yes	No
Indices	One index (0-1)	2 indices (0-1 and non-supplied water volume)	One index (0-1)

- **The number of hydraulically independent networks in the service:** it is indeed often the case. For instance in Lausanne there are 20 independent networks with different sizes. So a pipe that will have an Aquarel value equal to 1 will really have less importance than a pipe with a "1" value in a bigger network. It is then necessary to balance these HCI according to the size of the network or the water consumed in each network. The same issue occurs for RelNet. For Failnet-Reliab, the HCI is also proposed in term of non-supplied volume, which allows direct comparison with pipes from different networks.

2.5 CONCLUSIONS AND FUTURE PROSPECTS

The CARE-W project has included the testing and validation of failure forecasting models and hydraulic reliability models. These technical indicators are input parameters for the module CARE-W_ARP, where they are combined with impact and economic parameters to compute criteria useful for an Annual Rehabilitation Programme. Utilities need to record maintenance data that assign failures to pipe links. These databases can then be quickly used (with a minimum of 3 to 5 years' data) to give suitable results. Hydraulic reliability models additionally require pre-calibrated network models.

These two methods have been included in the suite of tools produced during the CARE-W project which includes planning an Annual Rehabilitation Programme (ARP), definition of long term rehabilitation strategy (LTP = Long Term planning)), control panel of performance indicators (PI) and tools to calculate technical indicators (FAIL and REL). This suite will provide water utilities assistance in choosing the best rehabilitation policy, according to their requirements, and consequently to define the data to be kept to implement this policy.

Other national and European research, working on new technical indicators, is in progress. These are mainly related to water losses and the impact of pipe condition on water quality, and will provide complementary indicators useful for Water Utilities in strengthening their rehabilitation planning process.

2.6 REFERENCES

EISENBEIS, P., LAFFRÉCHINE, K., LE GAUFFRE, P., LE GAT, Y., RØSTUM, J., TUHOVCAK, L., VALKOVIC, P. (2003) CARE-W: WP2 – Description and validation of Technical tools D4 – Report on the tests and validation of technical tools. CARE-W (Computer Aided Rehabilitation of Water networks), EU project under the 5th Framework Program, contract n°EVK1-CT-2000-00053. Cestas (F): Cemagref, October 2003, 54 p. + appendices (electronic document)

EISENBEIS, P., LE GAT, Y., LAFFRÉCHINE, K., LE GAUFFRE, P., KÖNIG, A., RØSTUM, J., TUHOVCAK, L., VALKOVIC, P., (2002a) CARE-W: WP2 – Description and validation of Technical tools D3 – Report on models description. CARE-W, EU project under the 5th Framework Programme, contract n°EVK1-CT-2000-00053. Cestas (F): Cemagref, August 2002.

EISENBEIS, P., POULTON, M., LAFFRÉCHINE, K., LE GAUFFRE, P., (2004) CARE-W: WP2 – Description and validation of Technical tools D5 – Guidelines for the use of technical tools. CARE-W, EU project under the 5th Framework Programme, contract n°EVK1-CT-2000-00053. Cestas (F): Cemagref, January 2004.

Eisenbeis, P., Werey, C., Laplaud, C., (2002b), L'enregistrement des défaillances pour améliorer la connaissance des réseaux d'eau potable, TSM, 2002, (6)

LE GAUFFRE, P., LAFFRÉCHINE, K., BAUR, R., DI FEDERICO, V., EISENBEIS, P., KØNIG, A., KOWALSKI, M, SÆGROV, S., TORTEROTOT, J.P., TUHOVCAK, L., WEREY, C., Criteria for the prioritisation of rehabilitation projects. D6 report. CARE-W (Computer Aided Rehabilitation of Water networks), EU project under the 5th Framework Program, contract n°EVK1-CT-2000-00053, June 2002.

3

Decision Support for Annual Rehabilitation

Pascal LeGauffre, INSA and Jean-Philippe Torterotot, Cemagref

3.1 DECISION SUPPORT FOR ANNUAL REHABILITATION PROGRAMMES

3.1.1 Background, basis for the work

CARE-W_ARP in the frame of CARE-W

In CARE-W, three complementary approaches are adopted for developing a framework of strategic, pro-active rehabilitation of water distribution systems. The first one (CARE-W_PI, § @) is the development of guidelines for continuous observation of the performance of the network and the utility. It requires the definition and evaluation of a set of performance indicators (concerning pipe failures, water losses, customer complaints, etc.) that are calculated on network level, zone level, or even on pipe level. Secondly, the development and the calculation of long-term strategies of rehabilitation is an approach that refers to the estimated service life of pipe types and to the efficiency of rehabilitation programmes (CARE-W_LTP, § @). Strategies are defined by the length of pipe types that will be replaced or renovated in the future.

The long-term-strategy module of CARE-W is designed for the forecast of future rehabilitation needs, and the benefits that can be expected from selected strategies by calculating the pros and cons in terms of indicators with or without rehabilitation. The third approach is a multi-criterion procedure for the evaluation and selection of pipes that should be candidates for rehabilitation. The CARE-W_ARP software is dedicated to the definition of Annual Rehabilitation Programmes. It proposes a multi-criterion representation of rehabilitation actions and a multi-criterion decision-support procedure.

CARE-W_ARP has been developed by INSAVALOR (URGC – Civil Engineering Research Unit) in cooperation with all the partners of the CARE-W project in a four-step process:

- Formulation of criteria assessing the impacts of the condition of each pipe of an asset stock,
- Development of multi-criterion procedures allowing the comparison of pipes and the definition of rehabilitation priorities,
- Development of the CARE-W_ARP software,
- Experiments and improvements of the software

Information required for the calculation of the decision criteria is derived from four main sources:

- Performance indicators (PIs): for example, criteria WQD (Water Quality Deficiency) and WLI (Water Losses Index) may be calculated by the CARE-W_PI tool at a zone or district level.
- Hydraulic Reliability software: calculation of criterion HCI (Hydraulic Criticality Index) requires hydraulic modelling and simulations provided by three alternative software tools: CARE-W_Aquarel, CARE-W_Failnet or CARE-W_Relnet.

- Failure prediction tools to supply predicated burst rates per pipe. Several ARP criteria assess the risk of failure of the assets. This is done by combining the consequence of failure (e.g. information about sensitive customers [SC], the number of people supplied by the pipe [NPS], the importance of the road under which the pipe has been installed [SR], etc.) and the probability of failure (e.g. predicted burst rate [PBR]) calculated in using the CARE-W failure forecasting tools (CARE-W_Fail_PHM or CARE-W_Fail_Poisson).

- Utility Databases: to supply information on the pipe environment, surface type, population supplied, and so on. All these data have to be obtained from data sources managed by the water utility or other utilities (roads & traffic, urban planning, etc.).

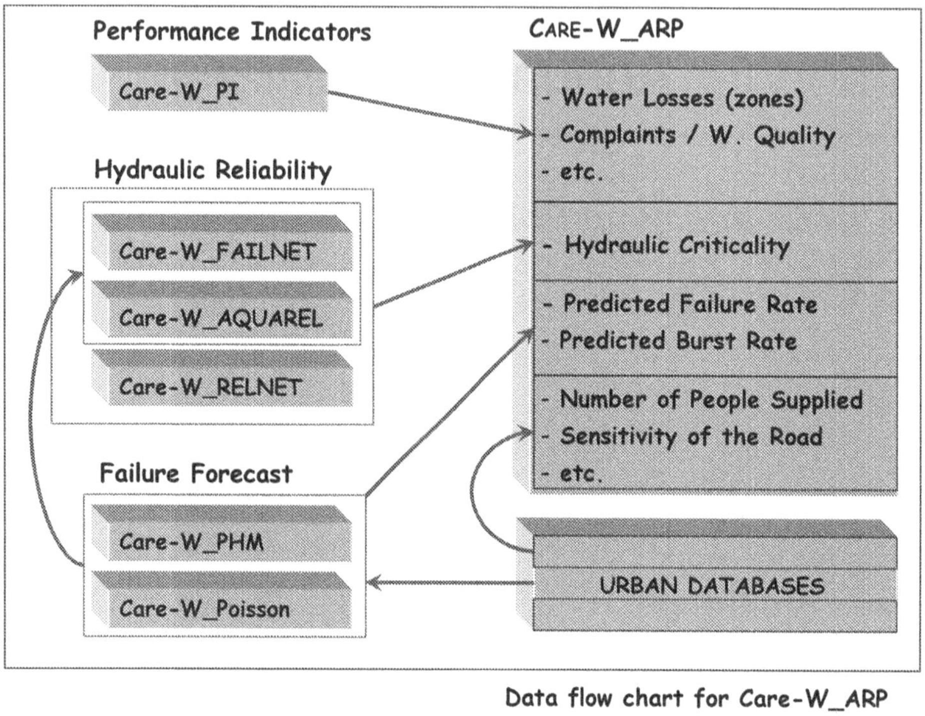

Figure 3.1. Data sources and data flow for CARE-W_ARP

Practices, and expectations expressed by end-users

Two questionnaire surveys (and detailed case studies) have been done in the frame of this work package. A preliminary survey (D6 report: Le Gauffre et al., 2002) has been used to identify decision criteria currently used by end-users, together with needs and expectations. These results have been used to set up our proposal of a multi-criterion model.

A second questionnaire survey and detailed case studies have been dedicated to "decision-making processes and socio-economic issues" (Torterotot et al., 2003). When trying to draw a synthesis on such a study, the first keyword is diversity of present decision-making processes:

- in objectives;
- in structure and methods;
- in information considered;
- in actors involved.

The model proposed for the generic decision-making process has allowed structuring the information to be collected and the analysis of it. The outcomes of the survey and of the case studies however don't allow to refine or adapt further this generic model: none of the components of this model shows to be similar for all situations considered. In average, long term planning is less formalised than short-term programming. The present real situations show diversified levels of integration between the two time scales and between the two types of decisions.

It appears that a decision support system, focusing on water pipe rehabilitation, should therefore be flexible to comply with present intangible rules and shares of responsibilities, to allow for a smooth implementation, and to accompany afterwards the evolution in processes which this DSS could facilitate or even foster. The needs and expectations expressed by end-users give evidence to that point.

The first challenge of the CARE-W decision support system will be a successful implementation in water utilities (directly or through consultants) for them to profit by its improving and integrated features. The second challenge could be to use this DSS – or its outputs - to develop an in-depth communication on rehabilitation between the utility and the first round of involved actors: city administrations and elected persons (city councils), road planners and managers, other utilities managers, customers, population.

3.1.2 Scientific methodology: principles applied for defining decision criteria and for developing a multi-criterion decision support

Principles for the construction of decision criteria

The objectives of network rehabilitation can be regarded from different "points of view":

- internal points of view refer to the view of the operator and are focussing on technical concerns and their corresponding costs: repair costs, water losses, rehabilitation costs, co-ordination with other utilities.

- external points of view refer to the view of the customer and affected third parties, such as road users, who in most cases will be represented by the responsible local authorities. These external points of view are corresponding to social concerns or social costs: damages and disruptions induced by bursts, hydraulic deficiencies and water interruptions, water quality deficiencies.

For water utilities, the decision problem is how to spend a given annual budget (potentially derived from a rehabilitation strategy) on the most efficient projects. So, the objective of the decision support module is expressed as the prioritisation of rehabilitation projects in a water supply network. It consists of the construction of criteria for the comparison of these rehabilitation projects, and the development of a multi-criterion procedure for the definition of priorities.

The decision-support for the proposal of annual rehabilitation programmes can follow one of these alternatives:

Prioritisation of pipes that are candidates for rehabilitation, by assessing performance deficiencies and ranking pipes according to their multiple deficiencies;

Prioritisation of possible rehabilitation measures (technique i applied to pipe j), by assessing performance improvements and rehabilitation costs and impacts, and ranking measures according to these multiple effects.

In the CARE-W decision support, the first formulation has been followed, due to the fact that the assessment of performance improvements after rehabilitation measures remains difficult for some particular aspects (e.g. reduction of water main breaks or water losses expected after each possible rehab technique).

The construction of the decision criteria follows two general principles:

Principle 1

A criterion represents the interest in improving an undesired situation or in avoiding performance deficiencies. This principle can be expressed in two ways:

- Corrective rehabilitation: identification and reduction of present deficiencies. These could be deficiencies identified on the pipe level by performance indicators on this scale, like frequency of water interruptions. This could be the contribution of a single pipe – identified by models or rules – to a particular problem in the network or a zone (e.g. red water problems or water losses) identified with performance indicators;

- Preventive rehabilitation: reduction of risks associated with the condition of the pipe (e.g. expected burst rate).

Principle 2

A criterion is a measure of a particular impact of the condition of the pipe. This impact can be identified by observation, by the calculation of a risk, or by assumptions.

In following these two principles, the information used in the formulation of criteria can be classified with respect to the data source. They relate either to the network performance or a single pipe's condition. Five types of information flows can be distinguished:

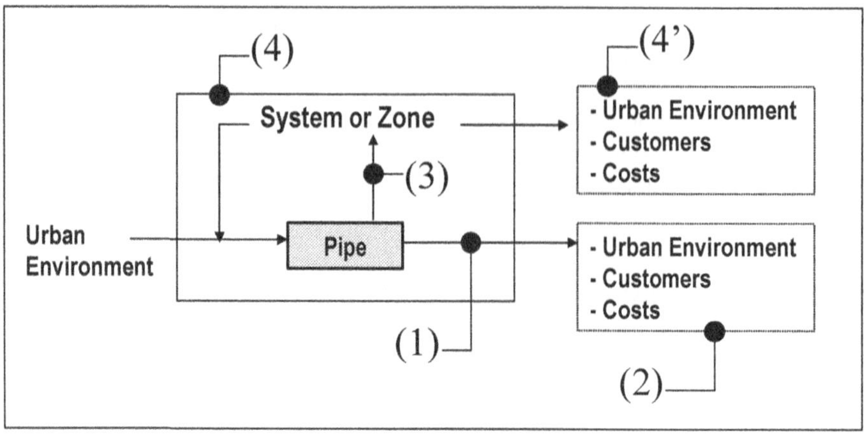

Figure 3.2. Information used in the formulation of decision criteria:
(1) calculation of potential local impacts, (2) observation of local impacts,
(3) assessment of impacts on system performances,
(4) observation of system deficiencies, (4') observation of impacts

Type 1: Calculation of potential local impacts (risks).

This approach is used, for example, for the assessment of the criteria "Damages due to flooding in housing areas" (DFH) or "Predicted Critical Water Interruptions" (PCWI).

Criterion DFH is the product of various sub-criteria, which represent:

- The risk associated with the condition of a pipe: sub-criterion PBR (Predicted Burst Rate),
- Aggravating factors assigned to the pipe: service pressure (P) and diameter (D),
- Aggravating factors assigned to the urban environment: sub-criterion "Intensity of flooding" (IF) is defined as an index between 0 and 1 that is evaluated by environmental characteristics of the pipe (slope, existence of basements, etc.),
- Vulnerability factors assigned to the urban environment: sub-criterion "Vulnerability to flooding in housing areas" (VFH) is an index between 0 and 1 and allows expressing the relative importance of values exposed to flooding in the case of a broken pipe (e.g. by land use patterns).

Criterion PCWI (predicted critical water interruptions) is calculated from the following sub-criteria: PBR, the Predicted Burst Rate, EDI, the Expected Duration of Interruption and SC, an index for the criticality of Sensitive Customers.

Type 2: Observation of local impacts.

A local concentration of customer complaints may be important information for the definition of rehabilitation priorities. The number of complaints related to water interruptions is a particular assessment of the impacts of a pipe failure rate.

Type 3: Assessment of the impacts of a deficient pipe on system performances.

The "Hydraulic Criticality Index" (HCI) is derived from hydraulic modelling with CARE-W_REL (hydraulic reliability analysis). This index measures the sum of the consequences of a pipe failure on all the nodes of the network.

Another example for the use of information of type 3 could be the criterion "Water quality deficiencies" (WQD). This criterion may combine results from rules or hydraulic modelling allowing to identify pipes potentially responsible for (information of type 3) and a performance indicator: red-water complaints observed on a particular zone (information of type 4').

Types 4 & 4': observation of system performance deficiencies.

This approach is used for example under the point of view "Water losses and relative costs" within the corresponding criterion "Water losses index" (WLI). In this criterion an observed leakage rate of a sub-network or a zone may be combined with characteristics of the single pipes and the corresponding marginal costs of water. A distinction can be made between technical observations carried out by the utility, such as leakage control measurements (information of type 4), and observation of deficiencies reported by the customer, such as complaints on water quality or pressure shortfall (information of type 4').

Principles for the development of a multi-criterion decision support

A criterion can be defined as a tool that allows the comparison of alternatives according to a particular "point of view". A criterion is a type of model allowing the establishment of preference relationships between alternatives.

Decision analysis is concerned in order to establish priorities between competing proposals. Its purpose is to provide a means of judging the economic, technical, social, environmental merits of alternatives. The question being raised may concern selection, allocation or ranking of actions. Adopting a multi-criterion paradigm means accepting that the decision is going to be a compromise between several conflicting objectives. There is no better a-priori method of approaching and dealing with these problems. It is useful in a first step to use representations by profile or stars. For the purposes of representation, criteria are supposed to be direct: i.e., the further from the origin, the stronger the preference. One main difficulty in these representations lies in choosing the scale.

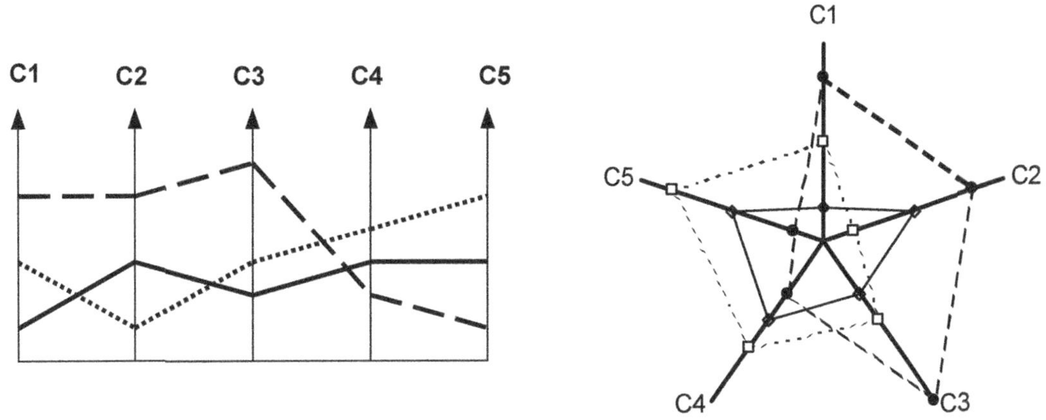

Figure 3.3. Case of 3 possible actions assessed on 5 criteria

In order to rank actions we have to deal with a critical and delicate problem within Multi-Criterion Decision Aid (MCDA): how to adequately assess the role-played by each criterion in a given decision problem, and how this translates into an appropriate weighting for each one? To be able to answer these questions it is necessary to clarify some aspects of methodological order. When the calculations or evaluations for the chosen indicators (an indicator is something that can be put a value on and be measured in some way, quantitatively or qualitatively) are done, they should be given weights. The next figure shows that this could be done on several levels, with different objectives and meanings.

Non-compensatory Models

3. All criteria are weighted in order to give the <u>relative importance</u> of each criterion involved. No aggregation is done.

2. The different indicators or sub-criteria are sorted into logic or natural groups and one criterion is calculated for each group

1. All indicators (or sub-criteria) are kept separately

Approach B

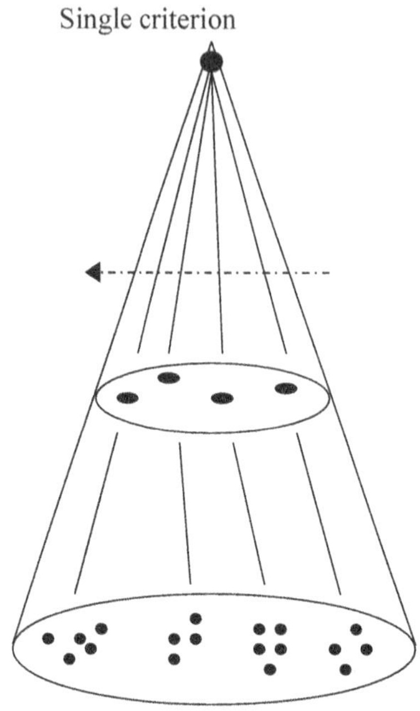

Compensatory / additive models

3. All indexes of level 2 are weighted in order to calculate a single criterion

2. The different indicators are sorted into logic or natural groups and one index is calculated for each group

1. All indicators are kept separately

Approach A

Approach A puts into practise a compensatory / additive procedure, approach B a non-compensatory procedure. Levels 1 and 2 of both situations are not fundamentally different, level 1 is a basic knowledge acquisition on the involved solution and level 2 groups natural belonging indicators (or sub-criteria) together. On level 3, all indicators (or criteria) are weighted for approaches A and B but the interpretation of weights is different.

Approach A – Compensatory / additive models

The methods used as checklists (Canter, 96) and multi-attribute-utility theory - MAUT, (Keeney & Raiffa, 76) are index or trade-off methods where individual attributes are transformed into units on a common notional scale, rating them in terms of their common importance, and then manipulating them mathematically in order to compute indices which allow the relative evaluation of alternatives. It can be assumed that different objectives can be expressed in a common denomination by means of trade-offs so that the loss in one objective can be evaluated against the gain in another (this idea of compensatory changes underlies the traditional cost-benefit analysis).

Compensatory / additive models seek to reduce option evaluation and selection to one in which each of the alternative options is classified using a single score which represents the attractiveness or utility of an option. The selection of a preferred plan is based upon these scores. Previous remark on the loss power discrimination is even stronger here (very different solutions may get a same index value).

These approaches seem relevant to the assessment of a solution or to the comparison of a proposition with a standard solution, or adapted in improving a proposition.

For comparison of a set of alternatives on different criteria, non-compensatory models seem more relevant.

Approach B – non compensatory models

The most commonly used methods are known as "outranking methods". The best-known outranking methods are the ELECTRE methods proposed by Bernard Roy (Université Paris Dauphine). Because it is a non compensatory decision-aid model the interpretation of weights is different than for a compensatory system such as MAUT, where they amount to being substitution rates, allowing differences in preferences, as they relate to different criteria, to be expressed on the same scale. Within ELECTRE the "weights" used are not constants of scale, but are simply expressing a measure of the relative importance of the criteria involved.

The ELECTRE method is probably the most widely used in the field of the environment (Rogers M., 2000). The method works by pair-wise comparison of actions, with the result expressed as an outranking relation able to disclose situations of indifference, weak preference, strict preference or incomparability between each action-pair. By means of outranking relations actions can be ranked directly or inversely. Direct ranking consists in extracting from the set of actions that or those which are preferred, repeating the operation with the remainder, and so on. Inverse ranking begins by extracting the least preferred action. The interest of having these two rankings lies in their respective effects on incomparable actions, which direct ranking tends to push back to bottom rank while inverse ranking tends to push them up to the top - thus, the difference between the two rank orders sheds light on incomparability. If a two-dimensional graph is used to display ranks obtained on the two types of ranking, with each action represented by a point on the graph, then stable actions will lie close to and actions with strong incomparability far from the diagonal (the closeness of a point to the top right hand corner of the diagram indicates a high ranking).

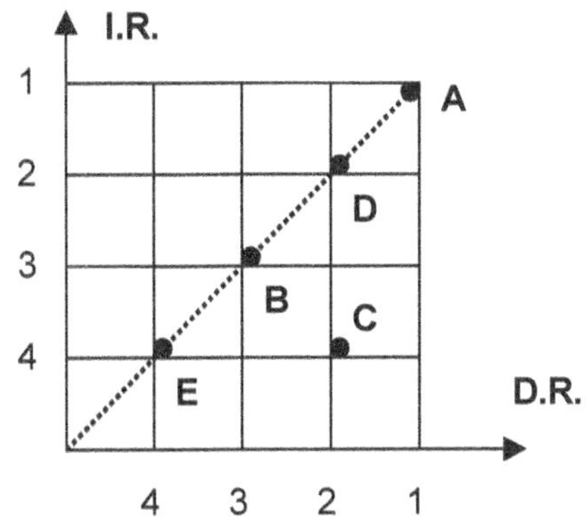

Figure 3.4. Direct ranking and indirect ranking, example with 5 actions

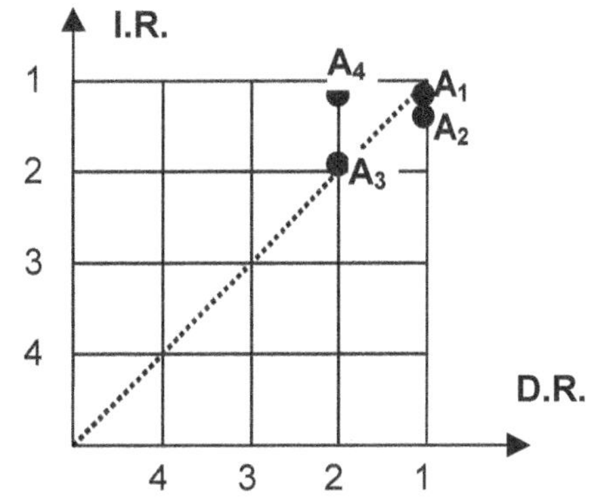

Figure 3.5. Strong consensus & good ranking for alternative A (four sets of weight)

Results need to be completed with sensitivity and robustness analyses. Sensitivity analysis examines the stability of results with regard to variation in the various parameters (performance and threshold values, etc.), and serves as a basis for robustness analysis. The above graph may usefully serve to visualise the results of the analysis. In particular, the positions of a given action can be displayed according to different weightings given on the criteria by different decision-makers, thus creating a cluster of points such that the smaller the cluster the greater the consensus. By comparing the locations of the various actions' clusters, a better idea of final ranking can be obtained.

ELECTRE (and other outranking methods, such as PROMETHEE) are in fact methods for aiding problem-design in decision-making, and serve for multi-criterion negotiation. They are above all a communication tool between the study team and the various actors taking part in the decision process.

In CARE-W_ARP, the ELECTRE TRI model is chosen for the prioritisation of rehabilitation projects. The selection of projects for a rehabilitation programme within a planning period does not require a hierarchical final order of all candidates. For the development of an ARP, rather the assignment of candidates to priority categories is of interest. Within ELECTRE TRI (Rogers *et al.*, 2000; Mousseau *et al.*, 2001) NC categories are defined in a hierarchical order by (NC-1) reference profiles, from category C_{NC}, for the best options, to category C_1, the category for options, which are "not relevant". For the assignment of an action (pipe) *a* to a category C_k, *a* is compared successively to the reference profiles The procedure is suitable in the case of a large number of candidates, with the aim of selecting just a bundle of actions for realisation but no need of an internal final hierarchy among these selected candidates.

3.1.2 Scientific results obtained

A multi-criterion model supporting the definition of annual rehab programmes

Criteria used in CARE-W_ARP are presented in the following table. Each criterion is defined as a combination of sub-criteria (or parameters) or is an index that can be customised by each end-user.

Table 3.1. Criteria definition

Short name - Long name	Definition	Units	Data source	Information needed
ARC - Annual Repair Costs	PFR × UCRp	No/100m/year euros	CARE-W_FAIL CARE-W_ARP / KB1	Failure history Repair cost / pipe type
WLI - Water Losses Index	Index ∈ [0,1]		CARE-W_ARP / KB5	PIs related to water losses
PWI - Predicted Water Interruptions	PBR × EDI × NPS	No/100m/year Hours Persons	CARE-W_FAIL CARE-W_ARP / KB7 Utility Database	Failure history Time to repair / pipe type No of persons supplied
PCWI - Predicted Critical Water Interruptions	PBR × EDI × SC	No/100m/year Hours Index	CARE-W_FAIL CARE-W_ARP / KB7 Utility Database	Failure history Time to repair / pipe type Sensitive customers
PFWI - Predicted Frequency of Water Interruptions	PBR × EDI × L/100	No/100m/year Hours m	CARE-W_FAIL CARE-W_ARP / KB7 Utility Database	Failure history Time to repair / pipe type Pipe length
HCI - Hydraulic Criticality Index	Index ∈ [0,1]		CARE-W_REL	Hydraulic data
DFH - Damages due to Flooding in Housing areas	PBR × D^2 × P × SFH or (IF×VFH)	No/100m/year mm^2 MPa Index	CARE-W_FAIL Utility Database Utility Database CARE-W_ARP / KB8	Failure history Diameter Water pressure Sensitivity of the urban area
DFI - Damages due to Flooding in Industrial or Commercial areas	PBR × D^2 × P × SFI or (IF×VFI)	No/100m/year mm^2 MPa Index	CARE-W_FAIL Utility Database Utility Database CARE-W_ARP / KB9	Failure history Diameter Water pressure Sensitivity of the urban area
DSM - Damages due to soil movements	PBR × D^2 × P × LS	No/100m/year mm^2 MPa Index	CARE-W_FAIL Utility Database Utility Database CARE-W_ARP / KB10	Failure history Diameter Water pressure Risk of landslide
DT - Traffic Disruptions	PFR × SR	No/100m/year Index	CARE-W_FAIL CARE-W_ARP / KB11	Failure history Sensitivity of the road
DDI - Damage and/or disruption on other Infrastructure	PBR × D^2 × P × SI	No/100m/year mm^2 MPa Index	CARE-W_FAIL Utility Database Utility Database CARE-W_ARP / KB10	Failure history Diameter Water pressure Sensitivity of infrastructure
WQD - contribution to Water Quality Deficiencies	Index ∈ [0,1]		CARE-W_ARP / KB6	PIs related to water quality issues (complaints, results of water analyses)
AUCR - Annual Unit Cost of Rehabilitation		Euros / 100m / year	CARE-W_ARP / KB3 CARE-W_ARP / KB16	Rehab costs Relevant rehab techniques
COS Co-ordination Score	Index ∈ [-1,1]		CARE-W_ARP / KB1	Work planned on other utilities (road, gas, etc.)

CARE-W_ARP has been designed to support the definition of annual rehabilitation programmes by taking into account all the possible impacts of the pipe condition and in allowing customised calculation rules: knowledge and information used for the calculation of criteria are provided by other modules of CARE-W or captured in "Knowledge bases" (KBs). These KBs are used to define the values of some criteria (COS, WLI, WQD) or the values of some parameters used in the calculation of criteria (UCRp, EDI, SR, etc.)

These KBs allow end-users to customise CARE-W_ARP to their individual circumstances (some examples are provided in the section dedicated to the Reggio Emilia case study).

Feasibility study

Table 2 presents some of the tests on an entire network or part of such a network that have been completed by the CARE-W partners: Reggio Emilia (R.E.), Lausanne (L), Oslo (O), Greater Lyon (G.L.), Brno (B) and Trondheim (T).

It appears that each proposed criterion may be relevant for at least one end-user. Information required for the calculation of the criteria is not always fully available, however the application of the CARE-W methodology provides a guideline for the improvement of the databases. On the other hand, for some cases, additional criteria have been defined and used by the utility: fire protection for instance.

Table 3.2. Criteria available in each experiment

	Water service:	R.E.	L.	O.	G. L.	B	T.
	Length studied (km):	632	800	8	1100	272	21
	Number of pipes:	2729	8000	215	19800	2000	279
	Status (oct. 2003):	done	done	done	→	→	→
	Criteria						
COS	Co-ordination score	✓			✓	✓	✓
ARC	Annual Repair Costs	✓	✓	✓	✓	✓	✓
WLI	Water Losses Index	✓					
PWI	Predicted Water Interruptions	✓		✓	✓	✓	✓
PCWI	Predicted Critical Water Inter.	✓			✓	✓	
PFWI	Predicted Freq. of Water Inter.	✓	✓	✓	✓	✓	✓
DFH	Flooding in Housing areas		✓		✓	✓	✓
DFI	... in Indust. or Comm. areas						
DSM	Damage due to Soil Movement				✓		
DT	Traffic Disruptions	✓	✓	✓	✓	✓	✓
DDI	Damage on other Infrastructure			✓			✓
WQD	Water Quality Deficiencies index	✓			✓		
HCI	Hydraulic Criticality Index		(✓)	✓		✓	✓
	Additional (user-defined) criteria	✓	✓,✓				✓

Prioritisation of rehab projects: multi-criterion procedures

Three procedures are suggested to support the definition of rehabilitation needs and/or rehabilitation priorities. Two methods have been implemented in CARE-W_ARP: (a) the ELECTRE TRI method, and (b) the Hotspots method. The alternative method, not elaborated here, is that of interactive elimination. This is the preferred method for another CARE-W partner (University of Dresden) and is used in their SPARENET software.

Hotspots: a hotspot is a pipe for which a particular criterion value exceeds a given threshold. The threshold has to be defined by the utility in order to express unacceptable performance deficiencies. An example is given in the section dedicated to the Reggio Emilia case study.

ELECTRE TRI: this procedure uses the multi-criterion profile of each pipe. The outranking approach of the ELECTRE Tri method (developed by LAMSADE - Université Paris Dauphine) allows the prioritisation of pipes by comparing each profile with reference profiles. In CARE-W_ARP three categories are defined in hierarchical order by two reference profiles (b1 & b2) from category C3 for the best options, to category C1, the category for options which are less acceptable (Figure 3.6).

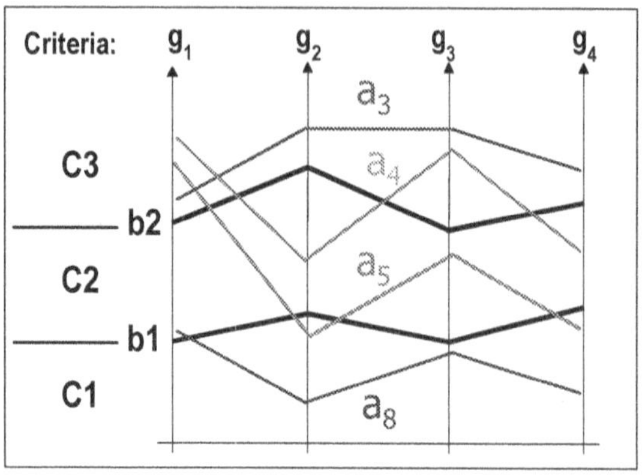

Figure 3.6. Two reference profiles (b1, b2) are defined for the “delimitation” of three categories. In this example, a3, a4, a5, a8 are four pipes (among a set of candidates) assessed according to 4 criteria (g1, g2, g3, g4) representing performance deficiencies

For the assignment of a pipe, *a,* to a category, *a* is compared successively to the two reference profiles. The comparison is used in two complementary procedures: an “optimistic” one and a “pessimistic” one. This double procedure applied to a given set of candidates leads to a result illustrated in Figure 3.7 and Figure 3.8.

Six categories of pipes can be distinguished:

- C33: pipes with highest profiles are assigned in C3 according to both the optimistic procedure and the pessimistic procedure. This sub-set of pipes is associated with the highest priority level.

- C32 (or C31): pipes assigned in these two groups have been assigned in C3 with one procedure and in C2 (or C1) with the second procedure. There is no consensus among criteria. For instance, pipe *a4* and reference *b2* (Figure 3.4) are judged *incomparable*, this leads to the final assignment: C32. Despite some low deficiency values, pipes assigned in C32 and C31 may be seen as good candidates for an annual rehabilitation programme (see an example in section 2.2.5).

- C22, C21 and C11: pipes assigned in these 3 groups are corresponding to low or moderate performance deficiencies.

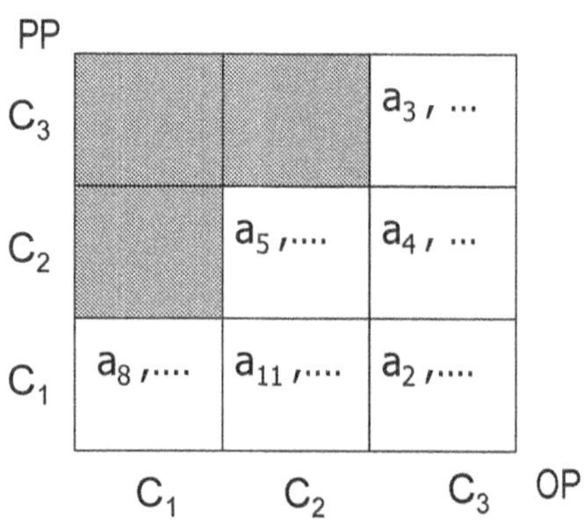

Figure 3.7. Candidates (a3, a4, a5, a8, etc.) are assigned to one of the 3 categories (C1, C2, C3) using both the optimistic (OP) and pessimistic (PP) procedures. Finally, six categories of pipes can be distinguished

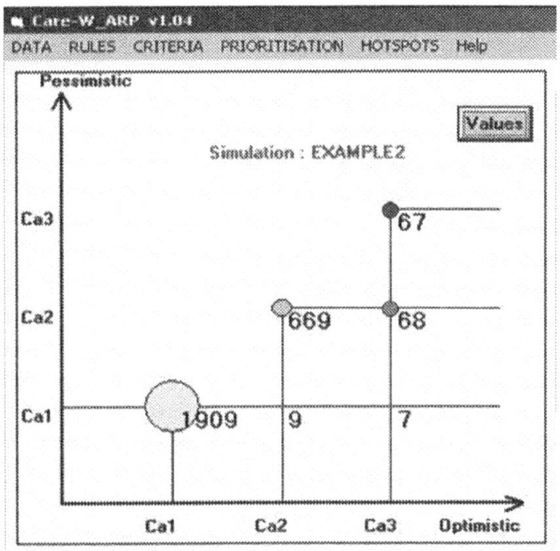

Figure 3.8. Typical results of a simulation with ELECTRE TRI. In this example, 67 pipes (out of 2729) are in C33, 68 in C32, 7 in C31, etc.

ELECTRE TRI: calculation rules

The outranking approaches of the ELECTRE methods develop a hierarchical order of candidates by comparison in pairs of all potential candidates in a concordance analysis (Rogers *et al.* 2000). Here, the ELECTRE TRI model is chosen for the prioritisation of rehabilitation projects. The selection of projects for a rehabilitation programme within a planning period does not require a hierarchical final order of all candidates. For the development of an ARP, rather the assignment of candidates to priority categories is of interest.

Within ELECTRE Tri (Rogers *et al.*, 2000; Mousseau *et al.*, 2001) NC categories are defined in a hierarchical order by (NC-1) reference profiles, from category C_{NC}, for the best options, to category C_1, the category for options that are "not so good". For the assignment of an action *a* to a category C_k, *a* is compared successively to reference profiles The procedure is suitable in the case of a large number of candidates, with the aim of selecting just a bundle of actions for realisation but no need of an internal final hierarchy among these selected candidates.

The comparison of each potential action *a* with each reference action *b* is done in using the performances of these two actions and four parameters associated with each criterion g_i (q_i, p_i, v_i : indifference- preference- and veto-thresholds, and w_i weight).

This comparison leads to one of these four possible conclusions:

1) *a* is preferred to *b* ($a > b$);
2) *a* and *b* are indifferent ($a\ I\ b$);
3) *b* is preferred to *a* ($b > a$);
4) *a* and *b* are incomparable ($a\ R\ b$)

The case of incomparability ($a\ R\ b$) can be explained with the two examples proposed in the following figure.

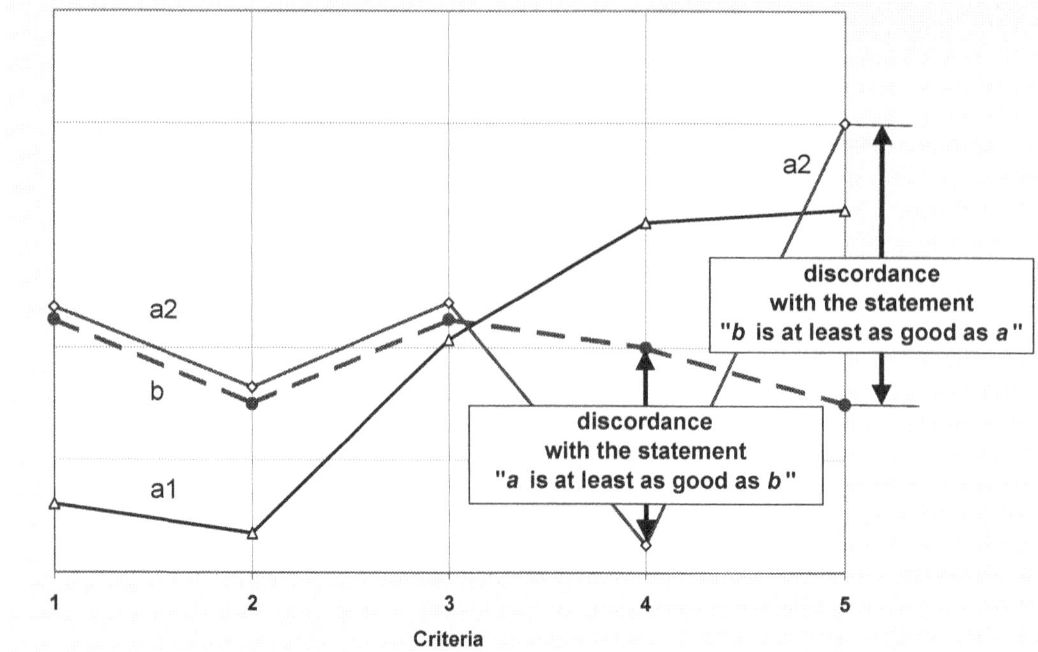

Figure 3.9. Illustration of the notion of incomparability

Depending on weights and thresholds, action *a1* and reference *b* may be judged incomparable: there are not enough reasons to prefer *a1* to *b*, and vice versa.

Depending on weights and thresholds, action *a2* and reference *b* may be judged incomparable: even if *a2* is "at least as good as" *b* for four criteria, criterion 4 could express a discordance that is strong enough to refute this statement; in the same time, criterion 5 could express a discordance strong enough to refute the statement *b* is "at least as good as" *a2*

The comparison of action *a* with the reference profiles is used in two procedures, which must be run through: an "optimistic" one and a "pessimistic" one. The two procedures use the same information but different rules to assign candidates.

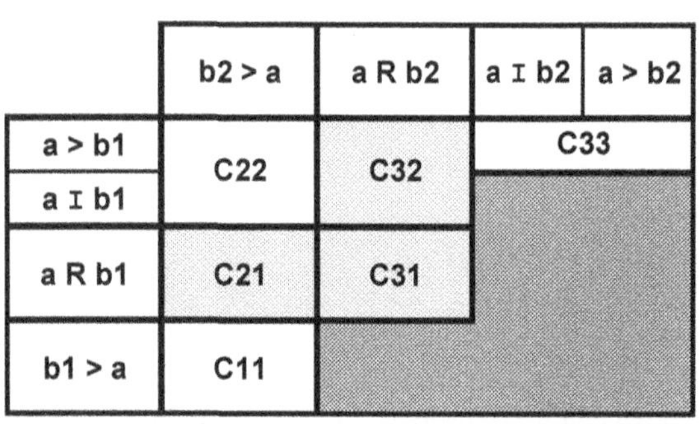

Figure 3.10. The six possible domains defined with the two assignment procedures (case of 2 references / 3 categories)

The cases where a candidate cannot clearly be assigned to one of the categories of priority by both the optimistic and the pessimistic assignment procedures are due to the incomparability with at least one of the reference profiles. For example, C32 means that action *a* is assigned to category C3 according to the optimistic procedure, and to C2 according to the pessimistic procedure. These two assignment results are different due to the fact that the corresponding actions are preferred to b_1 and are incomparable with b_2.

Implementation of ELECTRE TRI in CARE-W_ARP

For each criterion, **q** & **p** thresholds have to be defined by the user in order to represent imprecision associated with the calculation of the criterion and/or with the user's judgement.

These parameters determine the width Δ of a band. Profiles in this band will be indifferent with ***b2*** (see examples presented below). Δ is depending on λ (cutting level), which is a general parameter of ELECTRE TRI. Relevant values for λ are in the range [0.5, 0.75].

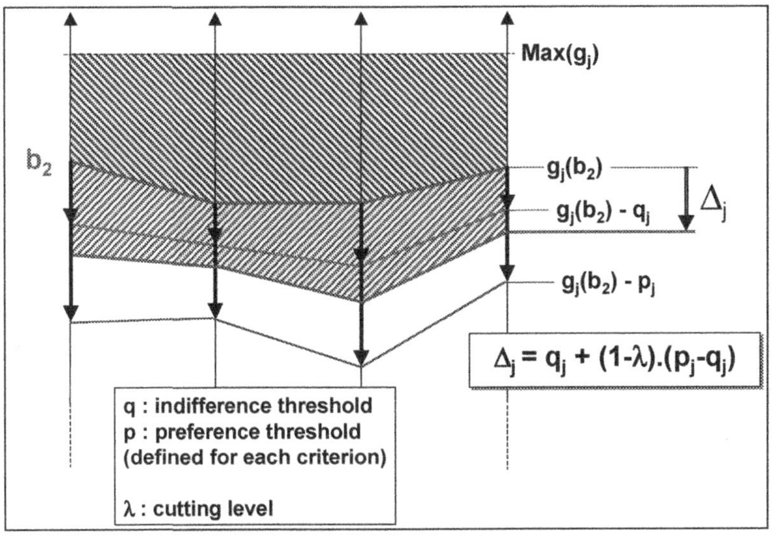

Figure 3.11. Example with 4 criteria

These thresholds are used by the ELECTRE TRI procedure when defining the relationship between an action ***a*** and a reference action ***b***. If profile ***a*** is under profile ***b2*** but close to it (band calculated with Δ) then the procedure will lead to the conclusion that **a is indifferent with b2**. Action ***a*** will be assigned in C33.

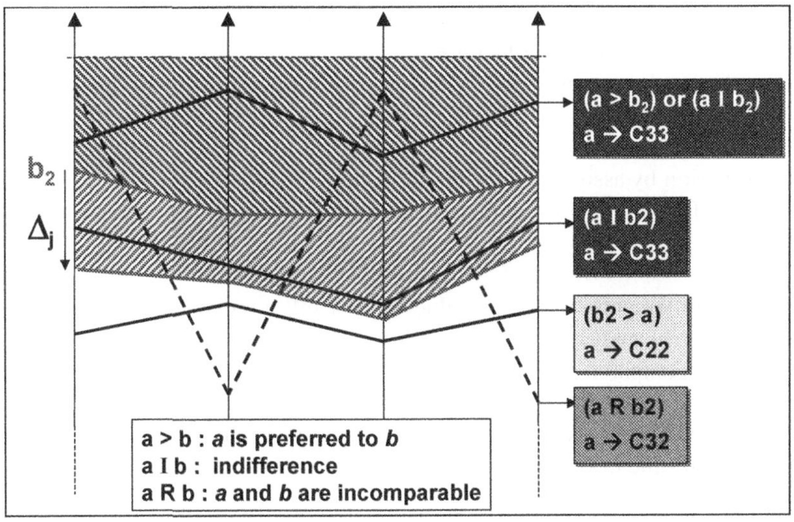

Figure 3.12. Relationship between a and b2: 4 possible cases (hypothesis: a is preferred to b1)

Default values for q and p are proposed by the software together with default values for g(b1) and g(b2). These values are calculated so that assignments lead to the following results:

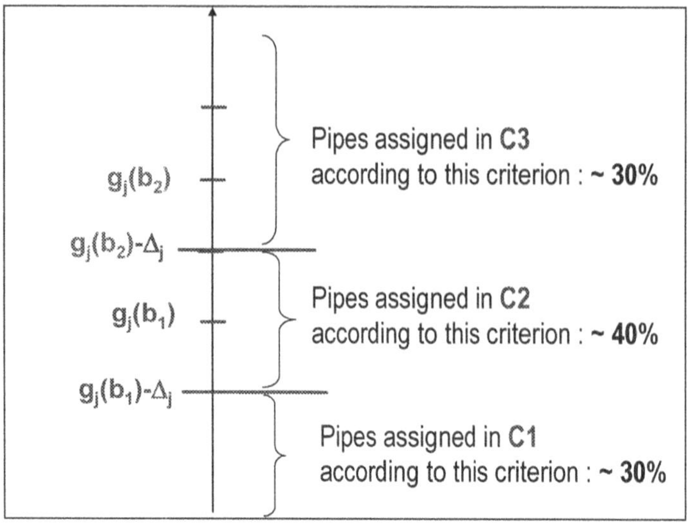

Figure 3.13. Rules used to define default values for g(b2), g(b1), q, p and v

These values may be modified. For instance, g(b2) may be raised in order to decrease the number of pipes assigned in C33. On the other hand, q and p may be reduced, if the imprecision is estimated to be less important.

We suggest to use a fixed ratio: p = 2.q
We also suggest to use the following rule for v and p: v > 3.p

How to apply the CARE-W_ARP module

Applying the ELECTRE TRI procedure in a new project is done as follows:

- Import raw data (ASCII text format). The input to CARE-W_ARP is of a pre-defined format and can be prepared using the CARE-W software's central data management facilities.
- Define rules for application of criteria (these rules are provided by the user by selecting or modifying what the ARP tool terms a 'knowledge base'.)
- Calculate criteria values using raw data and rules.
- Define the relative importance of each criterion by assigning a weighting factor,
- Modify suggested reference profiles and ELECTRE TRI simulation parameters, if necessary or as appropriate,
- Run an ELECTRE TRI simulation.

Several simulations may be run by modifying weights, reference values and/or other parameters. These results can be compared in CARE-W_ARP using sensitivity analysis.

Results obtained with ELECTRE TRI can be compared with results obtained with the "Hotspot" procedure (see next section).

Finally, results can be exported in a text file. Each pipe is assigned to one of the 6 possible categories (from C33 for the highest priority for rehabilitation, to C11 for the lowest priority).

3.1.3 Practical application: a full scale study performed with AGAC Reggio Emilia

Data collection and preparation of a database

Data have been obtained or derived from various sources (Figure 3.14):

- data provided by flow measurements and active leakage control (water losses),
- data provided by the customer service call centre (failures & complaints),
- data provided by other sources: customer consumption database, Yellow Pages, etc.

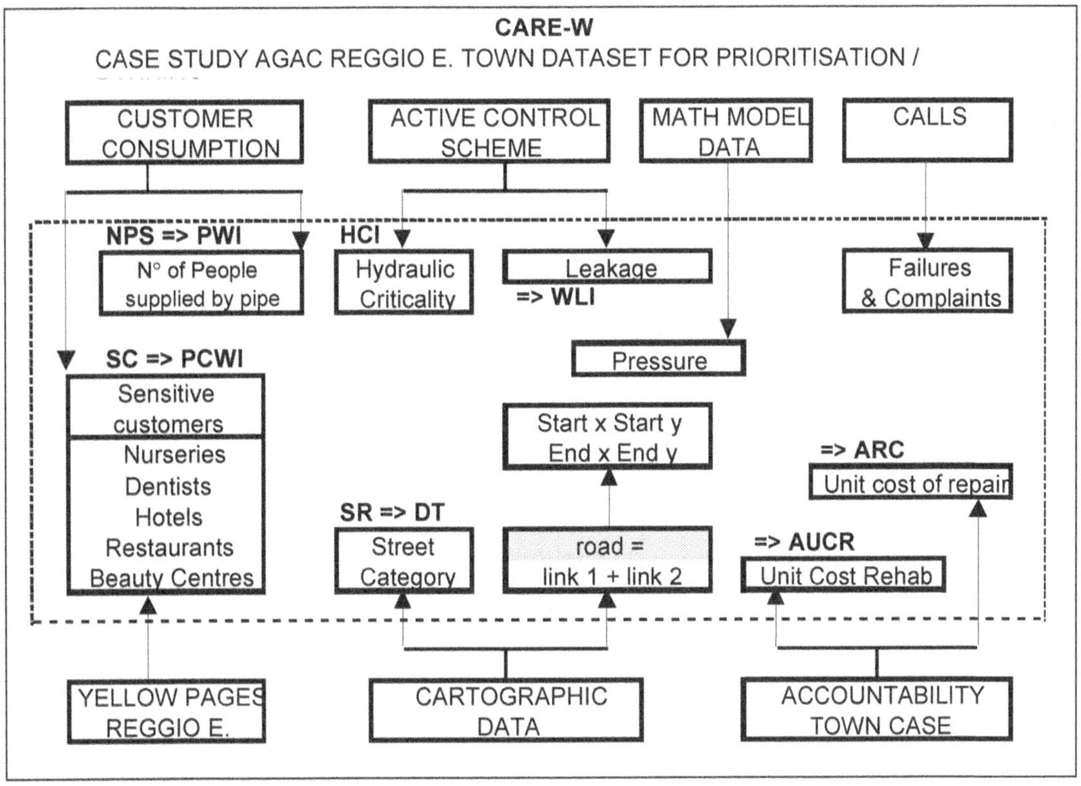

Figure 3.14. Data collecting at AGAC

Knowledge bases

Rules applicable for Reggio Emilia have been captured in several "knowledge bases". Examples are given below.

Code of category	Euros	Description
1	1100	Diam inf 100 mm
2	2200	Diam]100-200] mm
3	3200	Diam sup 200 mm
4	0	

Figure 3.15. Example of a knowledge base: KB2 is used to define pipe categories according to the Unit Cost of Repair (euros)

Code of category	Hours	Description
1	4	Diameter <= 110 & FIB
2	7	Diameter <=200 & FIB
3	10	Diameter >200 & FIB
4	3	Diameter <= 110 & other material
5	6	Diameter <=200 o& ther material
6	8	Diameter>200 & other material
7	0	

Figure 3.16. KB7: definition of EDI (hours) - Expected Duration of Interruption, used to calculate criteria PWI, PCWI and PFWI. Six cases have been defined by AGAC according to the diameter (3 classes) and the material (Asbestos Cement vs other material)

Calculation of PFR (predicted failure rate)

Predicted failure rates (PFR) were calculated using CARE-W_Poisson with failure records dating from 1994 to 2001. After several tests, the asset stock was divided into 18 pipe groups by using 3 available parameters: material (2 classes), diameter (3 classes), and type of road (3 classes).

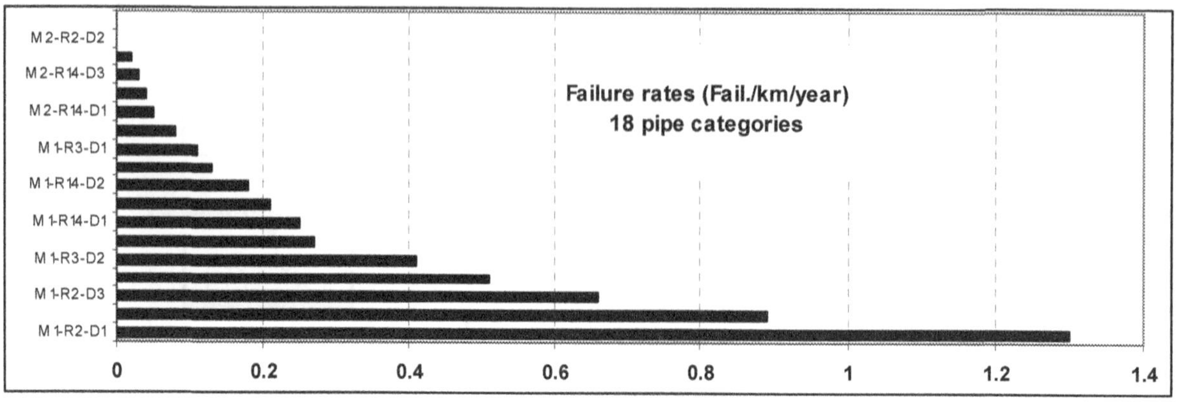

Figure 3.17. Failure rates calculated for 18 pipe categories (failures / km / year), defined by material, diameter and road type

Applying CARE-W_ARP: simulations done and sensitivity analysis

Eight criteria were calculated and available for the ranking process with ELECTRE TRI. Four simulations were performed using two sets of criterion weights combined with two sets of reference values (Table 3.3, Table 3.4).

Table 3.3. Two sets of weights (W1 and W2) used with ELECTRE TRI

Criteria	COS	ARC	WLI	PWI	PCWI	PFWI	DT	WQD
Set 1 (W1)	0	0.20	0.20	0.12	0.18	0	0.30	0
Set 2 (W2)	0	0.20	0	0.20	0.30	0	0.30	0

Table 3.4. Two sets (R1 and R2) of reference values used with ELECTRE TRI

Criteria	COS	ARC	WLI	PWI	PCWI	PFWI	DT	WQD
Maximum	-	2062	1.00	2947	3.281	-	0.750	-
PC_95%-ile	-	166	0.90	111	0.110	-	0.039	-
PC_5%-ile	-	0	0	0	0	-	0	-
Minimum	-	0	0	0	0	-	0	-
g(b2) in R1(default)	-	102	0.64	68	0.066	-	0.024	-
g(b2) in R2	-	170	0.80	110	0.100	-	0.040	-

Simulation S1 was performed with the initial set of reference values (default values calculated and suggested by the software). 215 out of the 2729 pipes were assigned to C33.

To reduce this number, values for reference b2 have been raised close to the 95th percentile of the corresponding criteria (Table 3.4). With this second set of reference parameters (R2), simulation S2 gives 49 pipes in C33, 105 pipes in C32, 30 pipes in C31 (Table 3.5).

Table 3.5. Number of pipes in the high or medium priority categories, from four simulations

Sets of reference values:	R1 (default values)	R2 (user defined)
Sets of weights:		
Set 1 (W1)	S1 simulation	S2 simulation
- No of pipes in C33	215 / 2729	49
- No of pipes in C32	89	105
- No of pipes in C31	0	30
Set 2 (W2)	S3 simulation	S4 simulation
- No of pipes in C33	226	56
- No of pipes in C32	62	95
- No of pipes in C31	0	40

A sensitivity analysis can be performed in order to compare results that have been obtained with different parameter sets. Figure 3.16 displays a table providing a comparison between simulation S4 (weights W2) and simulation S2 (weights W1): 45 pipes are assigned to C33 with both simulations, 1 pipe assigned to C33 in simulation S2 is assigned to C22 in simulation S4, and so on.

Synthesis table

REGGIOS4 ; ; REGGIOW2 ; REGGIOR2

Reference REGGIOS2 REGGIOW1 REGGIOR2

	Total	C33	C32	C31	C22	C21	C11
C33	49	45	3	0	1	0	0
C32	105	11	70	23	1	0	0
C31	30	0	14	16	0	0	0
C22	136	0	8	0	60	32	36
C21	34	0	0	1	5	17	11
C11	2375	0	0	0	10	8	2357

Figure 3.18. Results of a sensitivity analysis (S4 simulation versus S2 simulation)

Taking into account co-ordination with other utilities

Co-ordinating rehabilitation activities with other utilities' is an important driver for cost-efficient planning in many municipalities across Europe.

Criterion COS – Co-ordination score (Table 3.6) has not been considered when ranking pipes with the ELECTRE TRI procedure (its weight = 0). In a second step, this criterion has been used to select pipes among those assigned in C33, C32 or C31 with simulation 2 (Table 3.7).

Table 3.6. KB1: Evaluation of criterion COS: Co-ordination score

COS = + 1	Gas or sewer works planned in the same location
COS = + 0.5	Roadway rehabilitation
COS = - 1	Roadway has been rehabilitated during the last 10 years
COS = 0	Other cases

Table 3.7. Results of simulation 2 against criterion COS

COS = No of pipes assigned to …	- 1	0	+ 0.5	+ 1	Total
C33	2	37	6	4	49
C32	2	95	5	3	105
C31	0	28	2	0	30
C22	3	122	9	2	136
C21	0	30	1	3	34
C11	274	2009	42	50	2375
Total	281	2321	65	62	2729

In considering both the results of the ELECTRE TRI procedure and criterion COS, 20 pipes appear to be of high interest for the annual rehabilitation programme:

- 4 rehabilitation projects assigned to C33 may be co-ordinated with other utilities, 6 may be co-ordinated with road works,
- 3 rehabilitation projects assigned to C32 may be co-ordinated with other utilities and 5 may be co-ordinated with road works.
- At least 2 projects assigned to C31 (out of 30) could be seen as good candidates since road works are planned in the same location. The profile of one of these two pipes (pipe 942) is analysed in Figure 3.17.

On the other hand, one can notice that very few of the 281 pipes with COS = -1 corresponding to road works done during the last 10 years) are assigned to C33, C32 or C31. This is probably due to the fact that previous rehabilitation programmes have been defined in a close co-operation with other utilities: only 4 out of 281 pipes are in C33 or C32, 3 are in C22, and 274 are assigned to C11 (no or low performance deficiencies).

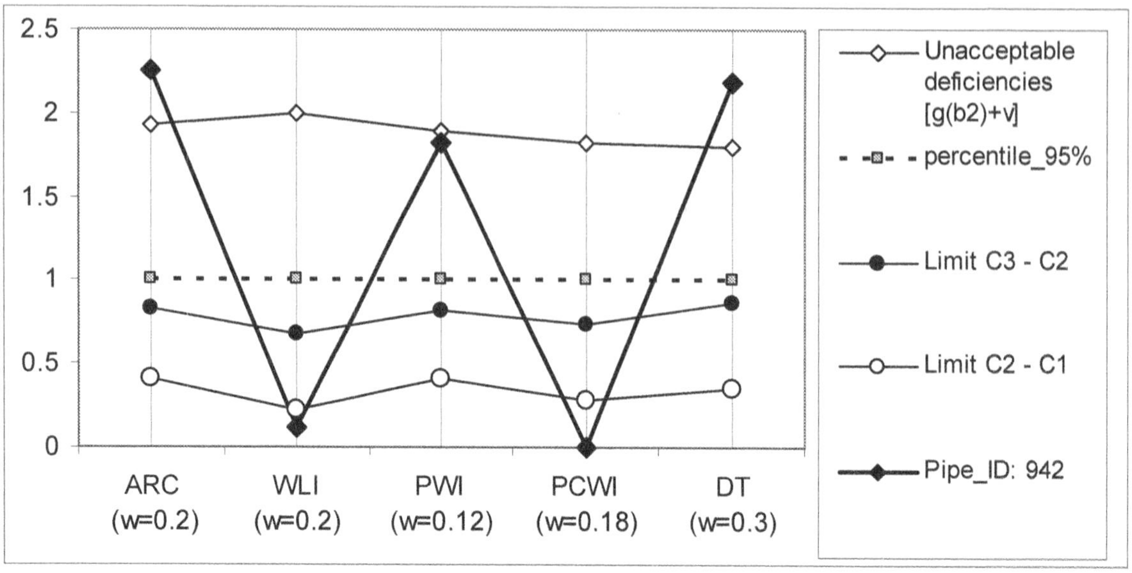

Figure 3.19. Multi-criterion profile of pipe 942 (scales have been standardised in using the 95th percentile). Despite some low deficiencies (low water losses, no critical interruptions) this pipe is assigned to C31 due to the fact that two impacts (ARC and DT) are exceeding a threshold

The ELECTRE TRI procedure & the Hotspots procedure

Results obtained with the ELECTRE TRI procedure can be compared to a selection procedure based on the "hotspot" notion. As an example, Table 3.8 and Figure 3.20 show the number of pipes identified as hotpots when defining a threshold for ARC, PWI or PCWI.

Table 3.8. Results of the ELECTRE TRI procedure and hotspots

Criteria	ARC	PWI	PCWI
Maximum	2062	2947	3.281
PC_95%	166	111	0.110
Chosen Threshold:	500	1000	0.500
% of hotspots:	0.77%	0.51%	0.88%
Assignment / simulation 2			
% of hotspots in C33	52%	78%	96%
% of hotspots in C32	38%	22%	4%
% of hotspots in C31	10%	0	0

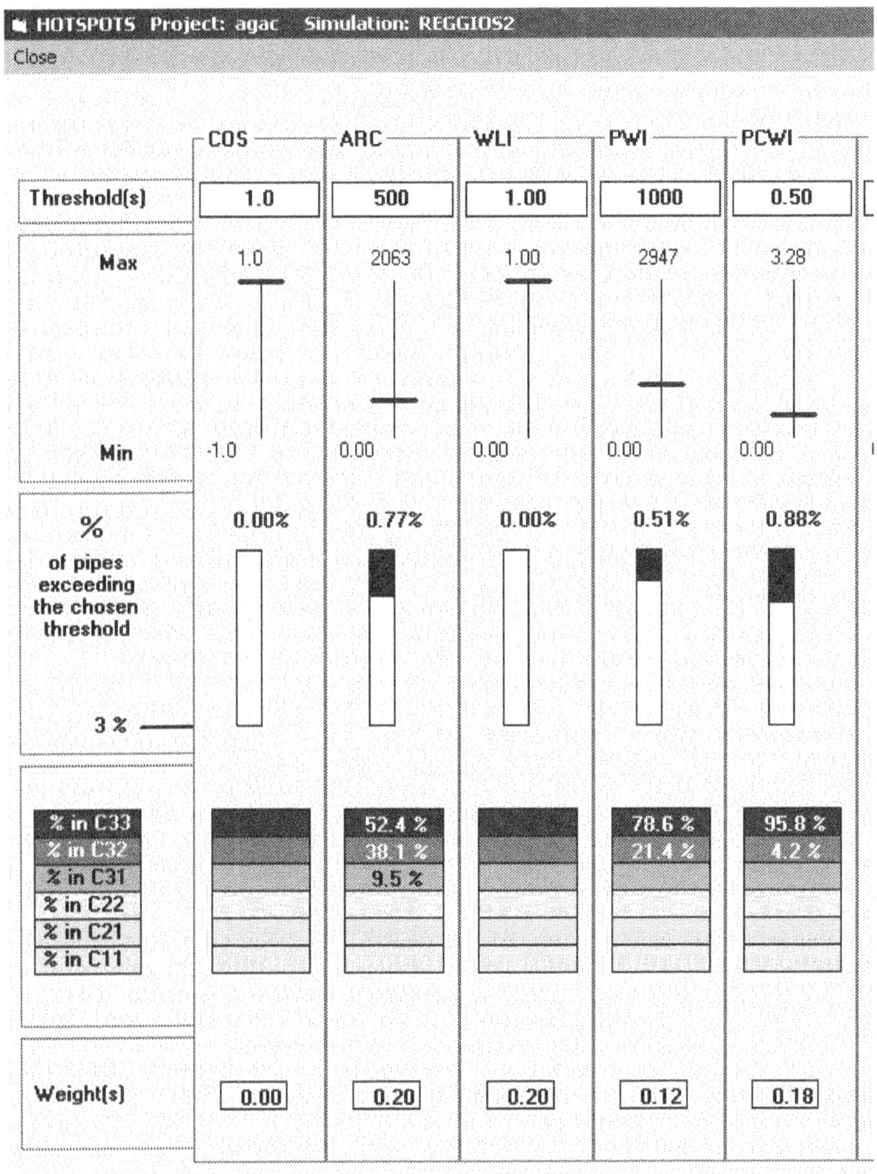

Assignment with ELECTRE TRI Simulation 2

Weights used in Simulation 2

Figure 3.20. Screenshot from CARE-W_ARP 1.05

For ARC, Annual Repair Cost (euro/100m/year), the chosen threshold (500) discriminates 0.77% of the asset stock. One can notice that all these pipes have been identified with the ELECTRE TRI procedure. 52% of these hotspots are assigned to C33, 38% to C32, and 10% to C31. Similar conclusions can be reached with criterion PWI or criterion PCWI.

These observations demonstrate the ability of the ELECTRE TRI procedure to process multi-criterion profiles without losing important information such as a single unacceptable performance deficiency.

Further experiments and research studies

A first round of experiments has been performed and allows validation of the principles and the implementation of the CARE-W_ARP module.

However, new questions arise and should therefore be addressed within complementary experiments and further research studies:

- Information for the calculation of criteria is more or less available / precise / reliable, depending on the end-user databases. This first issue has to be studied within further experiments. Data review and sensitivity analyses should allow evaluating the ability of the CARE-W_ARP procedure to deal with such data characteristics.
- CARE-W_ARP provides a ranking, which is based on the assessment of current and predicted impacts of the condition of each asset. Benefits of a rehab programme may be estimated in using these results. One critical point in this process is the prediction of pipe failures. Within WP2 (dedicated to failure prediction tools, § @)

this issue has been studied on several case studies: comparisons between predicted and observed failures have been performed, demonstrating and assessing the benefits of failure prediction tools (avoided failures). This evaluation has to be completed in assessing the impacts that could be avoided in using the ARP module.

- CARE-W_ARP supports the definition of annual rehab programs with two methods: the ELECTRE TRI procedure (ranking by assignment to 6 categories, in using 2 reference profiles) and the "Hotspots" procedure (unacceptable deficiencies defined by thresholds). The definition of these "reference profiles" and "thresholds" needs to be studied. One means is to perform sensitivity analyses within further experiments. These studies could lead to "recommendations" provided to end-users for "best practices" using decision support tools.

3.2 DECISION MAKING PROCESS AND SOCIOECONOMIC ISSUES

This chapter, presenting work Package 3.3b analyses current decision-making procedures and socio-economic issues as far as water networks rehabilitation is concerned. This task consisted in developing and consolidating a model on decision-making procedures, with a specific focus on socio-economic stakes, based on end-user interviews and on case studies on present decision-making processes.

3.2.1 Defining a conceptual model of decision-making for rehabilitation

When considering any organisation, decision-making processes can be divided into three types (Mintzberg, 1982):

- operational decisions;
- management decisions (co-ordination of operational decisions);
- strategic decisions (with an impact on organisations themselves).

They are bound to the structure, stakes and objectives of the organisation, as well as to existing information flows and decision-making tools. These processes can be analysed in terms of centralisation, formalisation, information (number of alternatives considered, data needed) and confrontation (Koopman and Pool, 1990). Water networks rehabilitation decisions can be considered in this general framework, and are concerned by the three types introduced above.

The study of real world decisions in organisations has led to questioning the implicit assumption of "optimisation". In practice, a decision within an organisation has to deal with several limitations and constraints:

- contradictory stakes and objectives (such as restricting rehabilitation expenses and restricting the impacts of coming network failures, for instance);
- level of available and manageable information (especially concerning forecasts and their intrinsic difficulty);
- time and resources;
- strength of habits;
- uncertainties;
- number of involved parties;
- decision criteria which cannot be compared directly (such as works costs, reduction of pipe burst probability, traffic disruptions).

For taking into account such limitations and constraints, was defined the "bounded rationality" decision model (Simon 1977, Simon 1983, Friedberg 1997). It accounts for facts like most decisions being taken without reviewing all alternatives and without checking whether optimal or not. The decision maker may make his decision as soon as identifying a choice which "satisfies him enough", given the objectives. By the way, spending too much time and resources for looking for a "better" option may not be optimal either, when considering both the decision process and the decision itself. This "bounded rationality" decision model focuses on a systemic description, with constraints, relations between involved parties, information flows.

Coming back to the case of water networks rehabilitation, a model of decision processes should include components such as: Problem identification and diagnosis, development / design of courses of action, selection / choice, review / approval. On the other hand, we have distinguished three different types of decision situations, which can be encountered for rehabilitation:

long term planning, in order to set for instance 1° generic goals to be pursued, 2° an interannual policy (levels of priority among techniques for intervention, among groups of pipes…), 3° orders of magnitude for annual budgets, and/or 4° priorities among areas of the networks; long term planning concerns the whole network; the time scale is typically 5 to 20 years;

short term programming, such decisions will refer either to goals set for rehabilitation at a long term planning stage, or to goals defined for this given year; short term programming concerns either the whole network, or may be done on each area inside the network; the time scale is typically 6 months to 5 years;

"urgent rehabilitation" or "non programmed" interventions, typically when new information reveals that a rehabilitation intervention may be necessary without waiting until the next programming period, or when a road is unexpectably opened (repairing is not meant here); such a decision refers to long term or to short term goals, and concerns a single pipe or a part of a street.

The surveys and case studies were based on the model described in the figure 3.21 below. They addressed among other issues: the existence of formal decision processes, involved parties, regulatory and contractual frameworks for rehabilitation, budgets, the description of the three types of decision situations.

3.2.2 Some results of the survey on present and actual decision-making process

An end-user survey was performed from December 2002 to February 2003, based on a questionnaire. The method considered for this survey was, for most cases, that project partners had an interview with the concerned end-user, over a time period that did not exceed 3 hours, for filling in the questionnaire.

Before showing synthetic and simplified results of the survey, it should be mentioned that the aim of the survey was to describe present decision making procedures, but no time to assess or judge them: the decision processes are the results of institutional situations, technical conditions, constraints, know-how, learning processes ... It would therefore make no sense to compare these processes between end-users, neither would it be sensible to have a definitive opinion based on such a superficial and rough description.

As can be seen below, the survey results show the diversity of situations among end-users when referring to rehabilitation planning or programming. They also give an inside into possible methodological needs.

The 14 end-users considered in the survey are the following ones: Bristol, Brno, Codigoro, Dresden, Ferrara, Lausanne, Lyon, Montijo (Serviços Municipalizados de Aguas e Saneamento de Montijo), Oeiras Amadora, Oslo, Reggio Emilia, Roubaix Tourcoing, Stuttgart, Trondheim.

The mandatory framework on rehabilitation

The basic compulsory or regulation framework is highly diversified from end-user to end-user, but in general we must stress on the fact that this framework is rather week concerning the definition of rehabilitation objectives or duties (most often this framework is expressed in general terms, or not expressed at all). For 4 cases, quantities of rehabilitation to be performed are defined and made compulsory (for 3), for 4 of them there are performance and level of service duties linked to rehabilitation, and 9 end-users are dealing with specific concerns on pipe materials to be changed at a given time horizon.

It can also be noted that quantitative rehabilitation duties concern, out of the Lausanne case, the cases of private utility managers (the formal definition of quantitative rehabilitation targets is not at all systematic when considering private managed utilities in general, for instance when considering the French situation). The English and Welsh situation is very specific compared with other parts of the European Union, with a sophisticated and formal planning and programming process.

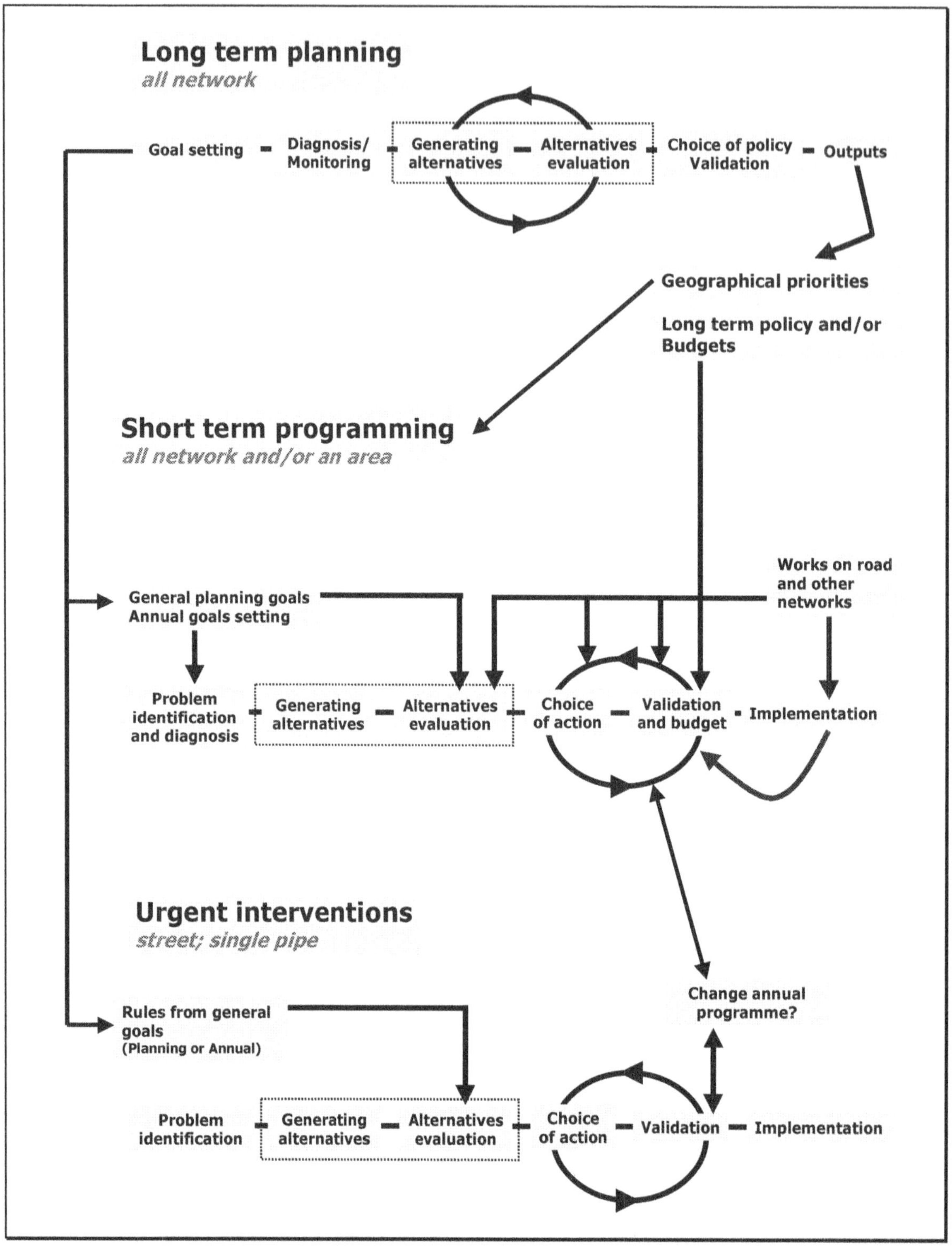

Figure 3.21: Process flowchart for rehabilitation of Water Network

Who proposes and who decides on rehabilitation: the network owner or the manager?

When the manager is different from the owner of the utility network, either public or private, he has an operational influence through proposition and often decision. But there is a formal control by the owner (in the concerned cases municipalities) through a contract, or through being a member of the board of the managing organisation. This raises the issue on the way the owner and the manager interact when considering rehabilitation (diagnosis of needs, negotiation, monitoring of performed rehabilitation actions): what is the information dealt with, how are stakes and objectives fixed?

Existence of formal decision procedures and some of their characteristics

The procedural and logical structures of rehabilitation decision making (planning, programming) are very diversified: for instance, budgets can be set a priori or following either a synthesis on needs or a prospective analysis. In the cases where there is no iterative process between technical analysis and budget definition, budget setting has most often the leading position. We can consider that an explicit and documented link between long-term planning and short-term programming appears in 6 cases.

Generic goals for rehabilitation, as expressed by the end-users for long term planning, are the following ones:

- life expectancy; average age of the network;
- compliance with management contract (quantitative targets for rehabilitation, or budgets);
- regulatory levels for service interruptions, pressure;
- voluntary standards or targets on levels of service;
- water quality;
- reduction of water losses, leakage;
- reduction of complaints;
- reduction of operating costs.

Time horizons considered for planning ranged from 3 to 25 years, for programming from 1 to 5 years.

In the majority of cases (11), once the annual rehabilitation budget is set it is compulsory or preferable to use it completely. If there is any obstacle on a programmed priority rehabilitation action, this can lead to implementing "second best" rehabilitation actions in order to use the annual budget.

Budgets for rehabilitation: sources of funds and financial trade-offs

The vast majority of surveyed utilities cover rehabilitation expenses only from the water bills, whether these expenses being investment or operation costs. In 4 cases, the municipal budget contributes to rehabilitation, and public subsidies do so in 3 cases. Water service customers are therefore, most often, direct beneficiaries of rehabilitation as well as direct payers for it.

In most cases the budget for rehabilitation is distinct from the first investment budget, and from the operation budget: if the budgets are not separate, it may enable the direct operational decision maker to do more easily trade-offs between projects and tasks of different natures: rehabilitate or repair, replace or maintain (from the answers to the survey, it can be thought also that "maintenance" has varying meanings, according to utilities, being either investment or operation from a technical as well as from an accounting point of view).

Interfaces with other actors

There is always a strong link with road works planning, apparently pipe rehabilitation or programming follows most often (completely, or partially) road programming (for 8 cases at least). It must be noticed that in the survey, the series of questions may have been understood ambiguously, therefore the picture of the dominant role of road planning (or the priority given to road works planning) should be considered with some care.

After road renewal, there is a ban of dig works in 7 cases (up to 5 years), and a possible ban (depending on roads) in 4 cases. The co-ordination with other utilities was reported in 5 cases. The integration or interfacing of programming for important works on different urban assets is an important issue. Decision support systems should allow for it explicitly.

Customers are primary stakeholders, as rehabilitation intends to secure service level and sustainability of assets management, but they have little direct influence on the decision process out the more or less formal taking into account of complaints (Grimble and Wellard, 1997). Let us remember that, as was shown above, they are the major payers for rehabilitation budgets. The growing concern on governance may oblige utility managers and owners to find ways to communicate with the public on the stakes of rehabilitation, in spite of the inherent difficulty for taking into account all aspects.

Decision making typology: levels of centralisation, formalisation, information and confrontation

A rough attempt was made to classify the decision procedures based on the survey. This classification considers degrees of centralisation, formalisation, information (use of data, documentation of the decision process) and confrontation. The results have not been validated with explicit and specific interviews, and should therefore be considered with much care.

There is an overall moderate level of confrontation. The "average" decision process concerning network rehabilitation is highly centralised out of the coordination with road works, rather formal, and the levels of documentation of the decision processes are very different from one end-user to another.

Needs and expectations expressed by end-users

Several end-users reported on recent fundamental evolutions in their practices, or in the goals of rehabilitation. The most commonly mentioned evolutions concern a greater investment on data and monitoring (GIS, works database). Other recent and current evolutions are more specific to each end-user:

- Bristol mentions a shift towards risk based approaches considering different requirements, shifting the scale from aggregated areas to asset level, after a decision making period driven by water quality, pressure, bursts; on the other hand, long term cost-benefit analysis of risk mitigation is a growing concern;
- Brno has elaborated a directive for rehabilitation planning;
- Oslo has fulfilled to some extend the goals on water losses reduction, and is now concentrating more on pipe condition, water quality;
- Lausanne: Medirel software has been existing since 2002; more responsibilities will be given to network managers for deciding rehabilitation works (diagnosis and budget);
- Roubaix-Tourcoing has raised the rehabilitation budgets, and considered a 3 years time period which enables more flexibility in programming; pipes are not only considered according to their condition, but also to their function and importance, as well as to their age; on the other hand, urbanisation and demand are evolving, which leads to considering pipe capacity changes;
- Stuttgart is improving experience-based prognosis models, for short and medium term rehabilitation needs;
- Trondheim is stressing on more planning, together with an increase in rehabilitation budgets, and is more focusing on risks.

The expectations and expressed needs of end-users may be summarised as follows:

- improving, refining choice and ranking methodology;
- assessing funding needs;
- linking long term and short term;
- improving monitoring / information management;
- supporting structural changes (change to inter-municipal utility, transfer of management responsibility);

through user-friendly and adaptable tools (regarding decision processes, objectives).

3.2.3 Detailed case-studies

Five cases where studied in more details, based on several interviews and contacts:

- Lausanne, in Switzerland, end-user of CARE-W project and surveyed above,
- Lyon urban area, in France, end-user of CARE-W project and surveyed above,
- Montijo, in Portugal, surveyed above,
- Oeiras – Amadora, in Portugal, end-user of CARE-W project and surveyed above,
- Strasbourg, in France (Communauté Urbaine de Strasbourg)

Each of these case studies is described with the following structure:

- Socio –demographic characterisation of the utility
- General features concerning decisions
    - Type of decision
    - Who?
    - How?
- Who does what?
- Budgets
- Specific models of decision
    - Urgent rehabilitation interventions
    - Short-term programming
    - Long term planning
- Praxis evolution during the 5 last and for the 5 next years

The figure below gives an example from the project case study of Strasbourg. It shows the way decision processes have been displayed. The dots represent different involved parties (persons, teams, departments).

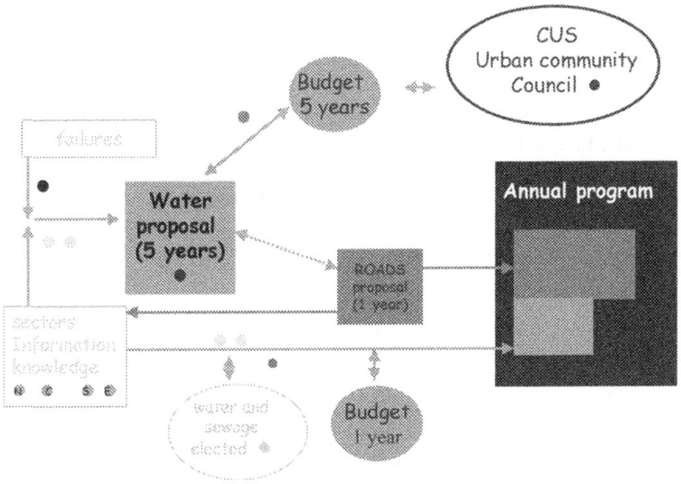

Figure 3.22: Example of practical decision process for water network rehabilitation

3.2.4 Concluding comments

When trying to draw a synthesis on such a study, the first keyword is diversity of present decision making processes:

- in objectives;
- in structure and methods;
- in information considered;
- in actors involved.

A decision support system, focusing on water pipe rehabilitation, should therefore be flexible to comply with present intangible rules and shares of responsibilities, to allow for a smooth implementation, and to accompany afterwards the evolution in processes which this DSS could facilitate or even foster. The needs and expectations expressed by end-users give evidence to that point.

Along this evident diversity, some common issues still characterise the majority of end-users' situations:

- no or little regulation framework for rehabilitation, with more often implicit and global objectives; quantitative objectives are set for the private managers among the end-users considered here in the sample;
- the budget for rehabilitation comes only or mostly from water bills;

- it is often compulsory or preferable to spend the whole annual budget allocated to rehabilitation (even if a priority action has to be postponed);
- when the network owner is not the manager, he is involved in decision, but sometimes in a very global and "superficial" way; the manager has a high influence on decisions;
- customers / citizens / inhabitants do usually not participate in decision, but claims and complaints play a significant role, and in some cases quantitative levels of service are addressed; the growing budget to be allocated to rehabilitation in many situations, as well as the growing concern on water price, may induce a change on that point;
- most decision processes are centralised, out of the link with road works programming.

The model proposed for the generic decision-making process has allowed the structuring of the information to be collected and the analysis of it. The outcomes of the survey and of the case studies however don't allow to refine or adapt further this generic model: none of the components of this model shows to be similar for all situations considered.

In average, long term planning is less formalised than short-term programming. The present real situations show diversified levels of integration between the two time scales and between the two types of decisions.

The first challenge of the CARE-W decision support system will be a successful implementation in water utilities (directly or through consultants) for them to profit by its improving and integrated features. The second challenge could be to use this DSS – or its outputs - to develop an in-depth communication on rehabilitation between the utility and the first round of involved actors: city administrations and elected persons (city councils), road planners and managers, other utilities managers, customers, population.

3.3 REFERENCES

Canter L.W., (1996), "Environmental Impact Assesment", Mc Graw Hill.

Keeney R.L. and Raiffa H., (1976), "Decision with Multiple Objectives : Preferences and Value Trade-offs", Willey.

Friedberg, E. (1997). Le pouvoil et la règle: dynamique de l'action organisèe. Paris, Seuil.

Grimble, R. and K. Wellard (1997). Stakeholder methodologies in natural resource management: a review of principles, contexts, experiences and opportunities. Agricultural Systems 55(2): 173-193.

Koopman, P.L. and Pool, J. (1990). Decision making in organizations. *International Review of Industrial and Organisational Psychology*, 5, 101-148.

Mangin J.C. and Miramond M. (eds), 2000, Proceedings "2nd International Conference on Decision Making in Urban and Civil Engineering", 20-21 November 2000, Lyon (France), vol.1, vol. 2 (1,270 p.).

Mintzberg, H (1982). Structure et dynamique des organisations (translated from "the structure of organizations" 1978). Paris, Les Editions d'Organisation.

LE GAUFFRE P., LAFFRÉCHINE K., BAUR R., DI FEDERICO V., EISENBEIS P., KØNIG A., KOWALSKI M, SÆGROV S.,TORTEROTOT J.P., TUHOVCAK L., WEREY C. (2002) Criteria for the prioritisation of rehabilitation projects. D6 report. CARE-W (Computer Aided Rehabilitation of Water networks), EU project under the 5th Framework Program, contract n°EVK1-CT-2000-00053, 72 p, June 2002.

LE GAUFFRE P., BAUR R., LAFFRÉCHINE K., MIRAMOND M. (2002) Survey of multi-criteria techniques and selection of relevant procedures. D7 report. CARE-W (Computer Aided Rehabilitation of Water networks), EU project under the 5th Framework Program, contract n°EVK1-CT-2000-00053, 31p. + Appendices, June 2002.

Mousseau, V., Figueira, J. and J.-Ph. Naux (2001) : Using assignment examples to infer weights for ELECTRE TRI method: Some experimental results. *European J. of Operational Research*, 130, 263-275

Rogers M.G. and Bruen M.P., (1998). "A new system for Weighting criteria within ELECTRE", European Journal of Operational Research, Vol. 107, N° 3.

Rogers, M., M. Bruen and L.-Y. Maystre (2000): *ELECTRE and decision support. Methods and applications in engineering and infrastructure investment*. 208 p. Kluwer Academic Publishing. Dordrecht (NL).

Roy B., 1996, "Multicriteria Methodology for Decision Aiding", Kluwer Academic Publishers.

Simos J., (1990), "Evaluer l'impact sur l'environnement." Presses Polytechniques et Universitaires Romandes, Lausanne.

Simon, H.A. (1983). Administration et processus de dècision (translated from "administrative behaviour: a study of decision-making processes in administrative organizations" 1976. Paris, Economica.

Simon, H.A. (1997). *Models of bounded rationality – vol.3: empirically grounded economic reason.* Cambridge (Massachusetts, USA), The M.I.T. Press.

TORTEROTOT J.PH. REBELO M., WEREY C., CRAVEIRO J., GUERRA R. (2003) Rehabilitation of water networks: decision-making processes and socio-economic issues. Report on task 3.3.b. CARE-W (Computer Aided Rehabilitation of Water networks), EU project under the 5th Framework Program, contract n°EVK1-CT-2000-00053, 66p. + Appendices, December 2003.

Vincke P., 1992, "Multicriteria Decision-aid", Wiley.

Software:

CARE-W_ARP, Version 1.05, INSAVALOR, 30-01-2004.

D8_Description_of_Care-W_ARP_software.HLP (july 2003) – updated feb. 2004

4

Strategic Planning and Investment

Raimund Herz, Rolf Baur and Ingo Kropp, Technical University Dresden

4.1 BACKGROUND

The ageing process of pipelines in drinking water distribution systems leads inevitably to a loss in performance, be it through bursts, leakage and hydraulic problems, through reduced water quality from internal pipe corrosion or through increased pumping energy to overcome bottlenecks. Thus, as more and more pipelines are approaching the end of their service life, water companies are facing rising rehabilitation needs in their water distribution networks. The common approach to answer this problem of network deterioration is to monitor a set of performance indicators and to take appropriate measures as soon as these indicators reach a critical value. The measures taken are subsequently monitored again with respect to the effects they have on the performance indicators (Baur/Herz 1999, Baur/Kropp/Herz 2003, Rajani/Kleiner 2002).

Such monitoring and controlling approaches need to be supplemented by a longer-term perspective for strategic network rehabilitation planning. Within CARE-W, strategic planning of rehabilitation investments was dedicated to working package 4. For strategic planning, the demand and supply side must be considered holistically; including trends in drinking water demand as well as technological and operational options for water supply in general and rehabilitation strategy in particular. Elements of the water supply system have to be presented in an aggregate form in close relation to key factors of the utility's controlling unit. As rehabilitation strategies are designed to fulfil a variety of more or less well defined objectives, before a rehabilitation programme is implemented, the effects of different options on the utility's objectives and performance indicators should be forecast and systematically compared with each other. Accordingly, the work was subdivided into the following three areas of research:

(1) Rehabilitation Scenarios (section 4.3)

Long-term scenarios help to evaluate future rehabilitation programmes. They have to be developed for local networks in a systematic way and specified for different options of rehabilitation up to the planning horizon. This may be a 5-50 years perspective. Some key factors are under the control of water utilities, others are difficult to predict in the long run. The creation of rehabilitation scenarios will serve as an important support for strategic decisions (CARE-W report D9).

(2) Rehabilitation Strategy Analysis (section 4.4)

Doing more or less rehabilitation, doing it sooner or later, doing it with more or less durable materials in a more ore less costly way, all this will have different effects in the short and long run. The rehabilitation strategy analysis will show the expected financial consequences of specific network rehabilitation programmes and their impacts on technical and economic performance indicators such as burst and leakage rates, maintenance cost, residual service lives, asset values and external cost (CARE-W report D10).

(3) Rehabilitation Strategy Evaluation (section 4.5)
A systematic approach must be taken to find the most appropriate strategy within a multitude of possible rehabilitation programmes. The rehabilitation strategy evaluation provides support for the selection of the most robust rehabilitation strategy that fits best within the predicted range of rehabilitation scenarios (CARE-W report D11).

4.2 FRAMEWORK

The methodology is mainly based on the well-known KANEW approach that was developed at Karlsruhe University (Herz, 1994) and internationally introduced by the research project "Quantifying Future Rehabilitation and Replacement Needs of Water Mains" funded by the AWWA Research Foundation (Deb et al, 1998; Herz 1996, 1998). The enhanced procedure (Herz, 2002) starts from a network inventory of pipe types by period of installation or rehabilitation (Fig.1). Pipe types are defined with respect to their ageing behaviour, mainly by material, capacity and corrosiveness. Failure statistics are used to support the definition of pipe categories and to estimate service lives, which can be represented by mathematical distributions. A cohort survival model for vintages of pipes calculates the mileage of pipes reaching the end of their service life. Due to uncertainties, this is done with lower and upper bounds of service life distributions. Within the resulting range of forecasts, decisions should be made on the various rehabilitation options for a specific medium term rehabilitation programme. The effects of alternative rehabilitation programme are simulated, and mostly transformed into monetary terms. Furthermore, long-term effects are simulated without the restrictions of specific programmes. Thus, several potential network rehabilitation programmes and strategies are explored in an interactive iterative process. This evokes the problem of selecting the best alternative. Thus, the existing KANEW framework was supplemented by two new features, the first one supporting the creation of utility background scenarios (Lipkow/Herz, 2002) and economic input data, and the second one for the evaluation of rehabilitation strategies, in order to find an optimal robust network rehabilitation strategy. The method for the evaluation of different rehabilitation strategies is the formalized balancing and ranking procedure (Strassert, 2000).

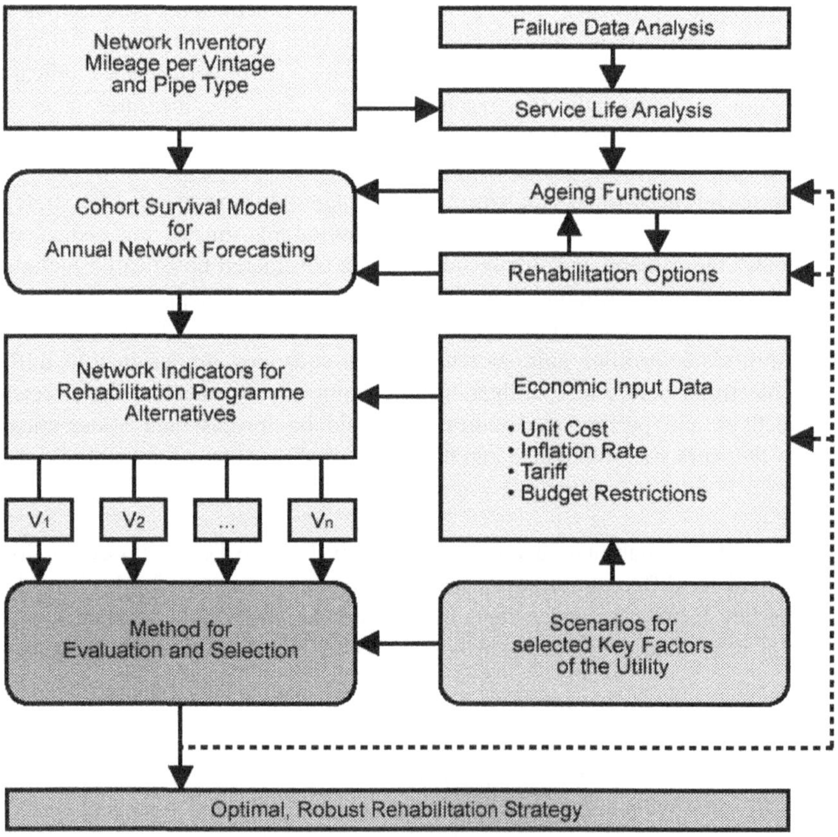

Figure 4.1. Extended *KANEW* framework for exploring rehab strategies

The *KANEW* approach is recommended by the German Association of Gas and Water Works (DVGW) in guidelines G401 / W401 as a method for long term forecast of rehabilitation needs of gas and water supply networks (DVGW 1997, 1999). The KANEW approach has been applied successfully to about 50 projects.

4.3 REHABILITATION SCENARIOS

4.3.1 Scenario Methodology

Scenario writing is a technique used for compiling enfolded images of the future from isolated perceptions of variances of development factors. The method originally comes from futurology and is not geared for planning. It was developed in the fifties in the USA and used as a soft forecasting method for nonlinear processes and unpredictable incidents. Later on it turned into a decision support tool. There are two main aims of these images of the future. They should have some probability of occurrence and they should be internally consistent and transparent, so they would be easy to communicate.

The scenario technique links quantitative data with qualitative information, evaluations and views to develop detailed and holistic descriptions of possible futures. Normally there are three scenarios to be created:

- The best case scenario – the positive extreme scenario, describing the most beneficial future development
- The worst case scenario – the negative extreme scenario, describing the most evil future development
- The trend scenario – the most probable scenario, describing the future development according to observed trends and common expectations.

Scenarios start from the present situation. The two extreme developments define the shape of a funnel opening up into the far future and containing possible future states (fig 4.2.). These future states are characterized by a set of consistent values of variables that are closely interrelated (Herz/Lipkow 2002). As time goes by, the funnel moves forwards.

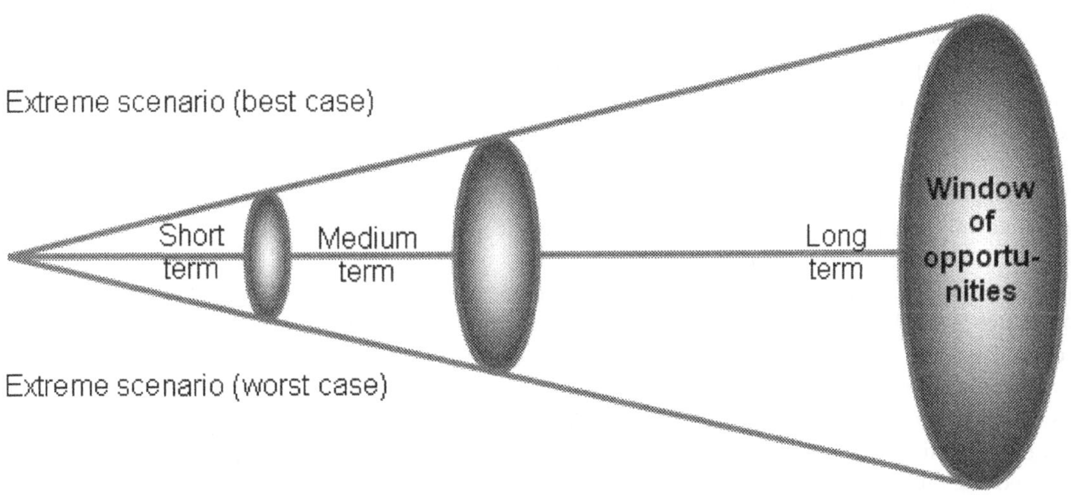

Figure 4.2. Scenario method – funnel into the future: "window of opportunities"

The development of scenarios is ruled by a fixed scheme:

1. *Problem/ system analysis*
   Starting points of each scenario are different problems that have to be solved and where controversial solutions or views exist. The first step is to define and delimit the system in its factual, spatial and time dimension.
2. *Impact analysis and key factor determination*
   The next step is to identify all impact spheres that are related to the problem in any way. These impact spheres are scanned in detail in order to find impact factors affecting the problem. Furthermore, measurement rules for these factors have to be defined either in quantitative (m^3/km•day) or qualitative (good, bad) terms. From all the factors affecting the problem, the most important ones have to be identified. By limiting the number of these so-called key factors to about a dozen factors, the subsequent analysis still is manageable.

	KF1	KF2	KF3	KF4	KF5	KF6	KF7	KF8	KF9
KF1									
KF2	**←**								
KF3	←	**←**							
KF4	←	←	0						
KF5	←	**↑**	←	↑					
KF6	←	↑	0	←	**↑**				
KF7	0	**←↑**	**←**	**↑**	0	**↑**			
KF8	↑	←	←	0	←	←	0		
KF9	↑	0	0	**↑**	↑	0	←↑	**←**	

Figure 4.3. Example of impact matrix

3. *Key factor impact analysis*
   All key factors are analysed in pairs in order to determine dependencies, interactions and relations between them. Impact matrices are used for keeping this difficult process transparent and manageable. These impact matrices must be internally consistent.
   The impact matrix in Figure 4.3 is an example. It shows the direction and intensity of relations between 9 key factors. There are n(n-1)/2 relationships (36 in this case), including zero-relations (0). The arrows are pointing from the affecting to the affected key factor. In case of an interaction, the cell contains two opposite arrows. Bold arrows indicate strong impacts. By counting the number of arrows pointing to or from key factors, active and passive key factors can be identified. Another useful category is whether relationships are proportional or inverse. This would normally be presented in a separate matrix.
4. *Projections of key factors into the future*
   The development of each single key factor, as defined by its measurement rule, has to be projected into the future for the best, worst and most probable case, on different time horizons (short term, medium term and long term). These projections of alternative future states have to be consistent and transparent. The consistency of these presumptions is validated by reference to the impact matrices.
5. *Scenario development and interpretation*
   This part – the so-called scenario writing – is the summit of the scenario technique. The consistent future projections of the key factors are now used to compile holistic future images illustrating in a vivid and evident manner possible future developments and their consequences.
6. *Conclusions, activity analysis and options of interaction*
   This final phase is connecting the problem analysis of the first step with the scenarios in order to develop strategies and to analyse the scope of activities needed to support favourable scenarios and prevent unwanted ones to occur. The main objective is to define a strategy that is robust with respect to potential future developments and to set up a catalogue of measures to be taken in the short and medium term.

4.3.2 Key Factors for Water Network Rehabilitation

For the establishment of rehabilitation scenarios 16 key factors (Table 4.1) were pre-defined by the CARE-W research partners (CARE-W Report D9). Most of them are context information on the water utility in accordance with the IWA PI project (Alegre et al., 2000) adopted for this particular purpose (Alegre, 2002). They refer to the total supply system or a few distinctive subsystems. The total key factor set is limited to 20 in order to minimize the complexity of the impact matrix, defined by the CARE-W experts, to support the writing of scenarios.

Table 4.1: List of pre-defined key factors

W1 Population supplied	E1 Domestic water tariff
W2 Total per capita consumption	E2 Industrial water tariff
W3 Residential per capita consumption	E3 Percentage of running costs
W4 Percentage of revenue water	E4 Inflation rate
W5 Yearly water production capacity	E5 Average per capita income
W6 Annual network extension	E6 Average number of persons per household
W7 Pipe length per capita	
R1 Network failure rate	Key factors R1, R2 and R3 are included in the set for plausibility control of the rehabilitation strategy forecasting results.
R2 Network leakage rate	
R3 Network rehabilitation rate	

4.4 SOFTWARE: THE REHAB SCENARIO WRITER

To support the scenario writing process the *Rehab Scenario Writer* software was developed. One of the difficulties in scenario writing is to account for the interrelationships between the various factors and to create consistent combinations of factor values that are not just wishful or fearful thinking on future states of the system. Therefore, the *Rehab Scenario Writer* checks on an impact matrix containing the direction and intensity of relationships between each pair of factors.

In case of inconsistent input values, the user is asked to reconsider and modify his first estimate. Thus, he is completely released from the first three steps of scenario development, and he can concentrate on the last steps of the scenario development: the key factor projections, analysis and scenario development, which is supported by the output of this software tool. The creative scenario writing itself still has to be done by hand and brain. The workflow model of the *Rehab Scenario Writer* software is shown below.

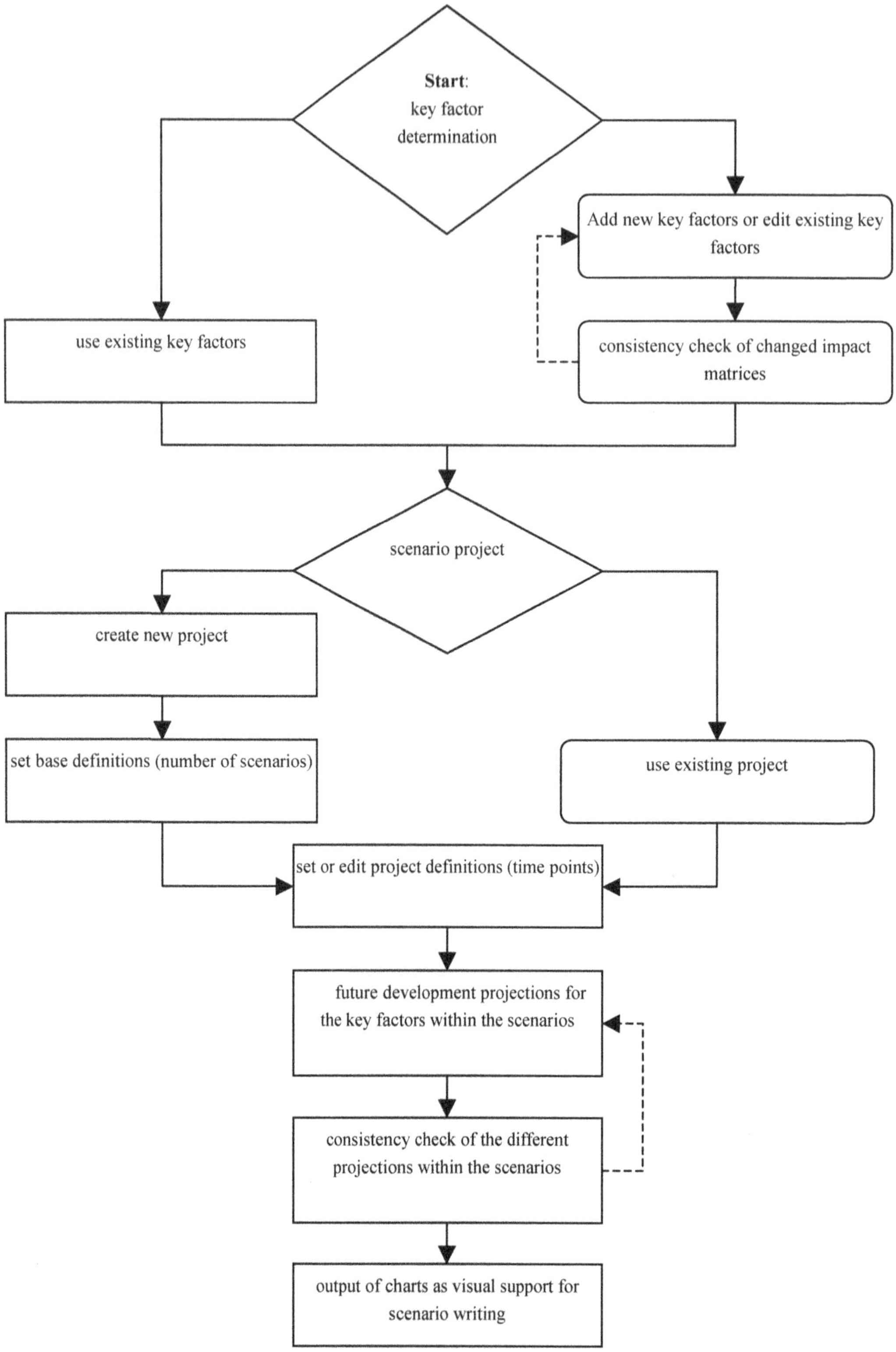

Figure 4.4. Workflow structure of the Rehab Scenario Writer software

4.5 REHABILITATION STRATEGY ANALYSIS

4.5.1 The Cohort Survival model

Ageing processes of pipes can be described mathematically as age-dependent probabilities of changing into worse conditions such as living or dead, sound or failed, function or malfunction. These rates of change, rates of death or rates of failure vary with the progression of age, depending on material and stress, and are functionally linked with probabilities of survival and life expectation, more generally formulated as the probability to be in a specific condition and hence to reach a corresponding age. This can be expressed in mathematical terms on the basis of Normal, Weibull, Gumbel and Herz distributions. There are four types of ageing functions that can be derived from these probability distributions (Fig. 4.5):

- a lifetime distribution function **f(x)**,
- a survival function **1 - F(x)**,
- a momentary failure/rehabilitation rate function **f(x) / (1-F(x))** and
- a residual lifetime expectancy function **R(x)**.

The lifetime distribution f(x) is a probability density function defined for positive values of age x. It is usually bell shaped or skewed to the left. By definition, "infant mortality" of infrastructure elements is excluded as a matter of guarantee. For types of infrastructure that have been laid decades ago, lifetime distributions can be derived from statistics on the age of particular elements at the rehabilitation date. However, due to financial and technological changes, these statistics of past cohorts do not necessarily hold for the future. For new types of infrastructure, there is very little empirical evidence how long they will last.

Survival functions are derived from the cumulative F(x) of the lifetime distribution. The percentage of elements of a cohort that reaches a particular age can be read from survival functions for specific types of infrastructure.

The rehabilitation rate of a cohort at a particular age is relating the elements that reach the end of their service life at a particular age to those that have survived so far. This rate is particularly useful in the simulation program because the stock of infrastructure is updated year by year. Again neglecting infant mortality of infrastructure elements, the rehabilitation rate starts from zero and gradually increases with age. A special feature of the Herz distribution (Herz, 1994, 1996) is that the rehabilitation rate is increasing progressively up to the median age, then turns into a degressive increase and finally approaches asymptotically a maximum rate. This means that the most resistant very old elements slow down their deterioration.

The residual lifetime expectancy starts from the mean value to decrease linearly at first, then degressively until it approaches a finite value close to zero, leaving some hope for survival. This results from the asymptotic behaviour of the rehabilitation rate in the Herz distribution.

The Herz distribution is implemented in the *KANEW* framework. It allows calibrating specific ageing functions for particular types of infrastructure by setting specific values for three ageing parameters:

- ageing factor a ,
- failure factor b and
- resistance time c .

Ageing factor, a, describes the smoothness of the starting phase of the ageing process. The larger this value is, the smoother the ageing process starts. For $a = 0$, the Herz distribution turns into the exponential distribution, which is not appropriate for describing the ageing process of infrastructure elements because the failure rate starts abruptly and remains constant all over the lifetime of the cohort.

Failure factor, b, is the final failure rate at very old age.

Up to resistance time, c, there is no rehabilitation, just spot repair in case.

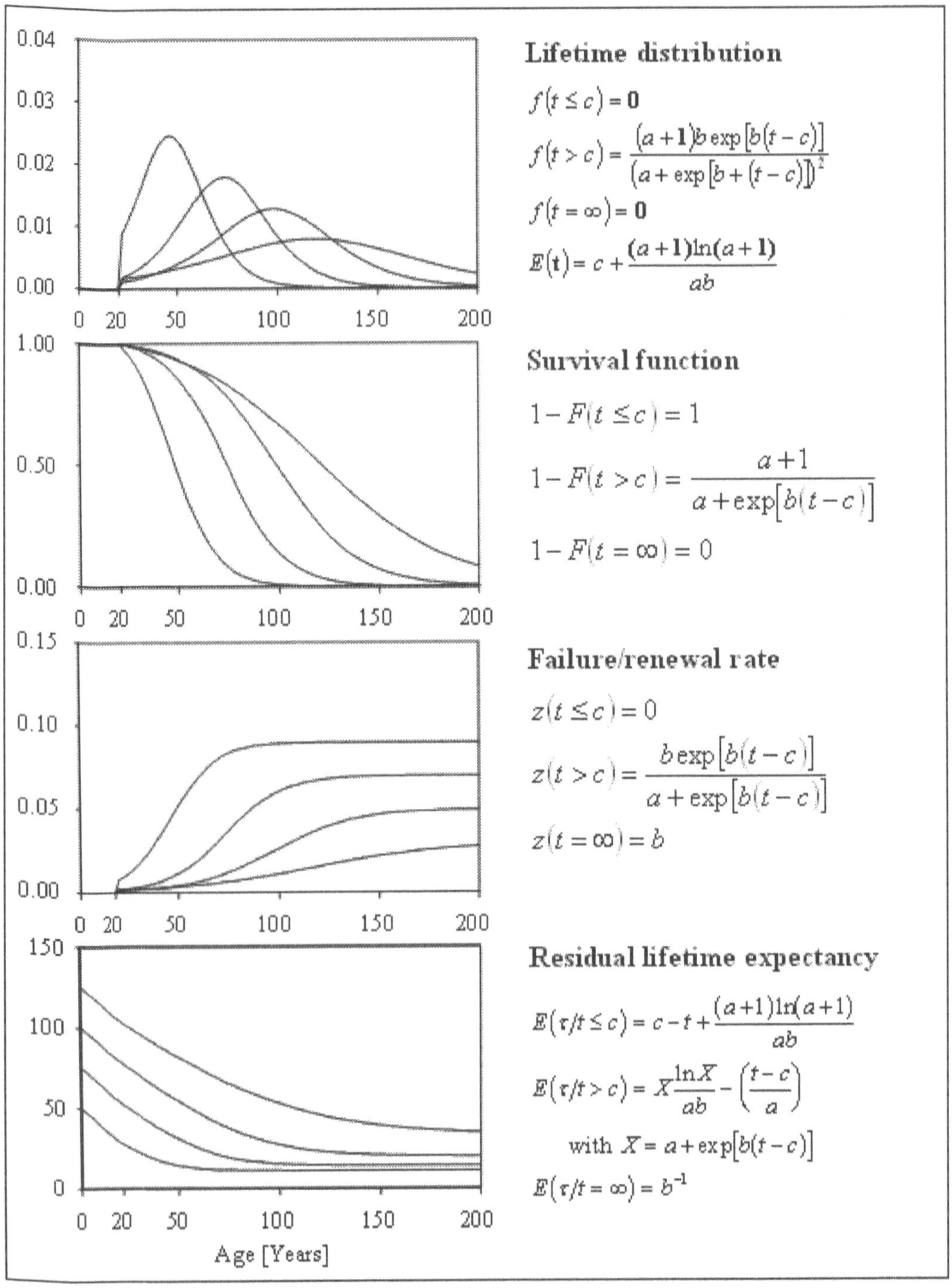

Figure 4.5. Ageing functions of the Herz distribution with a resistance time of 20 years and lifetime expectancies of 50, 75, 100 and 125 years

These parameters can be calibrated to fit a wide range of service life distributions for different types of infrastructure. Empirical frequency distributions may be derived from an analysis of failure and rehabilitation records or from expert knowledge processed in Delphi rounds (Kleiner/Rajani 1999). It is important to distinguish types of infrastructure with different ageing behaviour because, otherwise, the error of estimate would get too large. Even for a very specific type of pipe, such as ductile iron first generation (in West Germany laid in the 60s and early 70s of the last century), there is a considerable standard deviation of service lives which, in this case, results primarily from the degree of corrosive stress and generally also depends on the local rehabilitation policy and financial situation. Spot repair is always cheaper than replacement of the pipeline, at least in the short run. Thus, the cohort survival model should always be run for more or less optimistic assumptions on the service life of asset types and, by consequence, cover a range of future rehabilitation needs.

4.5.2 Predictable Performance Indicators

Performance Indicators (PI) are important instruments for monitoring and benchmarking the operation of the water utility. Whereas the *Performance Indicator Tool* provides an overview of the network performance in the past, the challenge within the long term strategic planning process is to facilitate the prediction of appropriate performance indicators with respect to a particular rehabilitation programme. There are two types of performance indicators in the PI-System of CARE-W (see chapter 1) that can be predicted. One type is related to rehabilitation work (Table 4.2) and the other to impacts of specific rehabilitation programmes on the network performance (Table 4.3).

Table 4.2. PI's related to rehabilitation work

Rehabilitation indicators		
Mains rehabilitation	%/year	IWA code: Op15
mains relining	%/year	IWA code: Op16
replaced or renewed mains	%/year	IWA code: Op17
replaced or renewed valves	%/year	IWA code: Op18
Service connection rehabilitation	%/year	IWA code: Op19

Table 4.3. PI's related to impacts of rehabilitation programmes on network performance

	PHYSICAL INDICATORS		
1	Mains residual service life	years	Code: Ph15
2	Service connection residual service life	years	Code: Ph15a
	OPERATIONAL INDICATORS		
3	Mains failures	No./100km/year	IWA code: Op26
4	pipe failures	No./100km/year	Code: Op26a
5	joint failures	No./100km/year	Code: Op26b
6	valves failures	No./100km/year	Code: Op26c
7	service connection insertion point failures	No./100km/year	Code: Op26d
8	Service connection failures	No./1000 connections/year	IWA code: Op27
9	Real losses	l/connection/day	IWA code: Op24
10	Infrastructure leakage index	-	IWA code: Op25
	QUALITY OF SERVICE INDICATORS (QoS)		
11	Pressure of supply adequacy	%	IWA code: QS9
12	Water interruptions	%	IWA code: QS11
13	Water taste	%	Code: QS16a
14	Water colour	%	Code: QS16b

Within the cohort survival model, all indicators related to rehabilitation work are considered. This is done with respect to different ageing functions for the particular asset types (pipe types, service connections and valves). According to the specific ageing behaviour of the asset types one can estimate the appropriate residual service lives (Table 4.3: physical indicators 1 and 2) with the cohort survival model. During the rehabilitation strategy analysis all operational indicators (Table 4.3) are predicted. The "Infrastructure index", standing for the percentage of water losses, can be calculated by simply using the PI "Water produced" (m^3/year, Code A7) as reference from the *PI tool* (chapter 1).

Since the rehabilitation strategy analysis does not forecast on the single pipe level, Quality of Service (QoS) indicators (11, 13 and 14) cannot be forecast yet, because they are closely related to problems in a local area. The QoS indicator "Water interruptions" can be predicted using the average duration of the interruption when a failure occurs.

4.6 STRATEGIC FAILURE PREDICTION

The prediction of failures is an essential element within the rehabilitation strategy analysis. Based on the future development of the number of failures, several indicators such as future repair costs, savings from reduced repair, balance between investments and savings etc. are calculated.

The current CARE-W rehabilitation strategy analysis uses a combined approach for failure forecasting. With respect to the existing stock, future failures are calculated using the current failure rate and a yearly increase rate for each asset type. These data can be derived from the failure analysis and forecasting tools in work package 2 (chapter 2). For rehabilitated assets a rehabilitation efficiency factor is used considering the failure rate of a pipe prior to rehabilitation in relation to the average network failure rate. Thus, the overall network failure rate is calculated by taking the annual increase rate into account in combination with a reduction of the number of future failures with respect to the rehabilitation efficiency.

Empirical research was carried out on the order of magnitude of the rehabilitation efficiency factor by some of the CARE-W partners. SINTEF has analysed rehabilitation work in the past in the Norwegian cities Oslo and

Trondheim. WRc's study is based on the rehabilitation work of the associated end user in Bristol Water, UK. The findings of these studies are summarised in Table 4.4.

Table 4.4. Rehabilitation efficiency factors for renovated and for replaced pipes

	Rehabilitation efficiency factors for:			
	Renovated pipes		Replaced pipes	
	average	range	average	range
SINTEF, Norway	6.3	4.1 – 11.1	9.5	4.2 – 24.1
WRc, England	1.2	0.3 – 2.2	3.3	0.0 – 10.5

From the results of these studies the following conclusions can be drawn:

1. The range of rehabilitation efficiency is quite large. Thus, it is necessary to analyse the specific local situation of the utility when determining the rehabilitation efficiency factor.
2. As stated in the internal report from WRc, high failure rates may quite often be not the main reason for rehabilitation. Therefore the *Rehab Strategy Manager* should consider rehabilitation work due to other reasons than high failure rates (e.g. water quality problems, advantage of co-ordinated replacement of pipes with other public work), where failure reduction plays a minor role. Thus, a smaller rehabilitation efficiency factor would have to be chosen.
3. The ratio between the efficiency factor of replaced and renovated pipes is about 1.5. This could be implemented in the software as a standard (which may be changed by the user).
4. The average rehabilitation efficiency tends to decrease year by year because decreasing failure rates due to the improvement of the network condition imply a decreasing rehabilitation efficiency. At present, a decreasing trend-function of rehabilitation efficiency is not included, but will be considered in the futurere after studying the characteristics of that efficiency curve.

4.7 SOFTWARE: THE REHAB STRATEGY MANAGER

To perform the rehabilitation strategy analysis the *Rehab Strategy Manager* software was developed as a support tool for the simulation of the effects of specific network rehabilitation programmes.

The *Rehab Strategy Manager* is the newly enhanced version of *KANEW* with interfaces to the *Rehab Scenario Writer* (see chapter 4.3) and the *Rehab Strategy Evaluator* (see chapter 4.5). The software allows not only to forecast the rehabilitation needs of user specified types of pipes (as in the first AWWARF version), it also allows to define user specified annual rehabilitation programmes for any time period, and to calculate the effects within the programme period and beyond. Thus, it can account for the effects of zero or postponed rehabilitation in total or for specific types of pipelines.

The main features of the *Rehab Strategy Manager* are described in the next sections of this chapter.

Starting the project

With input for names of the company, project, network specification etc.

Defining asset types and survival functions

Whereas assets such as pipes can be easily categorized according to materials, diameter, pressure, type of joints, anti-corrosive protection, and origin of production, it should be kept in mind that both, length or numbers and period of construction and rehabilitation respectively are needed to run the model. Furthermore, as assets should be distinguished with respect to their ageing behaviour, empirical analyses of failures and rehabilitation over age should be carried out first, before defining asset categories.

The parameters of the survival function of each asset type are automatically determined either from putting in mean, standard deviation and resistance time, or the age reached by 100 % (resistance time), 50 % (median service life) and the last (most durable) 10 % of assets. These figures are input by the user with a lower and upper bound, corresponding to shorter or longer service life assumptions. In general, it is recommended not to define asset types with a fraction of less than 1 % of all assets because such small categories will hardly improve the forecasting result.

The more asset categories there are, the more difficult it is to make consistent estimates of their life spans. For new materials, there is very little empirical evidence on how long these assets will last. Failure statistics often do not refer to the age of specific asset types, nor do rehabilitation statistics. Therefore, at least for some asset categories, intelligent guesses have to be made about the age that would be reached by certain percentages of them.

Survival curves from utilities provide some orientation and might help to re-adjust first round estimates in a second Delphi round. However, the local situation should always been taken into account and lower and upper bounds should always been given.

Realistic estimates of the assets' life span are the most crucial model input. They must be based on local experience and statistics of failures and rehabilitation activities in the past and should reflect past and future rehabilitation policies. The average service life of a particular asset type might be in the range of 60 to 80 years. How will the life spans be distributed around these two median points of age? Bell shaped or skewed distributions with the longer tail towards the older ages are more realistic than an equal distribution over a smaller or larger period of time.

It is recommended to estimate, in addition to the median service life, two points of age that would be reached by two specific percentages of each particular type of water mains. First, the age up to which no rehabilitation work would be done, just spot-repair in case of failure. This 100% point of the survival curve represents the so-called resistance time, which is one of the parameters of the survival function. Resistance time is large, maybe 30 to 50 years, where fire brigade policy prevails, with lots of spot-repair instead of preventive rehabilitation. Spot-repairs do prolong the assets' life spans, increasing its expectancies in general. The second point on the survival curve, besides the average expected service life, should be far beyond the median age and present the most resistant part of each particular asset type.

It is recommended to specify the age interval when the last 10% of that particular asset type will still be in service. Of course this interval will be larger than the intervals for the median age or the resistance time.

Creating or importing the network data base

The software requires input data on the mileage or number of assets and on their ageing behavior. Starting point of the ageing process is the year or period of installation or rehabilitation. So it must be given for each piece of asset that still is in service. The input of this data set can be done either manually or by importing files in csv format (csv: comma separated values).

Forecasting rehab needs and effects

There are several options to choose:

- running the forecasting model without constraints
- running the forecasting model with a specified network rehabilitation programme and programme period
  - without dynamic investment calculus
  - with dynamic investment calculus

For the first option, the following output charts can be generated:

- Inventory of assets by category and installation year
- Distribution of assets by installation year
- Age distribution of assets
- Cumulative age distribution of assets
- Survival functions for asset categories based on long life expectancies
- Survival functions for asset categories based on short life expectancies
- Average age and residual life expectancies of asset categories
- Cumulative residual life distribution of assets, based on short and long life expectancies for asset categories
- Length or number of assets to rehabilitate, based on long life expectancies of asset categories
- Length or number of assets to rehabilitate, based on short life expectancies of asset categories
- Annual rehabilitation rates for asset categories, based on long life expectancies
- Annual rehabilitation rates for asset categories, based on short life expectancies
- Annual rehabilitation rates for assets, based on short, median and long life expectancies for asset categories.

The second option requires the specification of a network rehabilitation programme.

Specifying a rehab program

Utilities can choose from a wide range of options to find an appropriate long-range network rehabilitation strategy, such as the extent of

- rehabilitation versus spot repair
- replacement versus relining
- no dig versus open trench replacement
- replacement alone versus with other public work

- metallic versus other pipe materials
- area wide versus section wise rehabilitation.

Any combination of these elements will have specific effects on the performance of the system in the short and long range. Based on the estimated annual rehabilitation needs of different assets, medium term rehabilitation programmes can be specified for the chosen rehab program period.

Forecasting the effects under the restrictions of the rehabilitation programme period

In the rehabilitation programme period, the network rehabilitation rates are fixed and it is only the effects that have to be forecast. Beyond the rehabilitation programme period, network rehabilitation rates also have to be forecast, under more or less optimistic assumptions on the assets' service life expectancies. This is done by re-assigning unfulfilled rehabilitation needs into the stock and calculating the accumulated rehabilitation needs at the end of the rehabilitation programme period, again under more or less optimistic assumptions on the assets' service life expectancies. The long-term effects are simulated without restraints, as if the system were rehabilitated at the rates forecast by the cohort survival model.

Future failure and leakage rates are calculated as described before. Thus, not only rehab investment costs but also the number of repairs and the volume of leakage saved by a particular network rehabilitation programme can be forecast and evaluated.

Transforming the effects into monetary terms

Specific unit costs and prices can be inserted on a particular mask. The figures depend on the local price level. Usually direct costs are taken. Indirect costs are accounted for by a social cost factor, which can be chosen accordingly. A distinction is made between total costs and the percentage of fixed and variable costs, which is important for valuing water losses. Variable costs are taken as a proxy for marginal costs of water saving. Inflation rates can be chosen according to the prospects of the national economy.

Balancing costs and benefits from rehab programs

Because rehab investments come first and benefits later, accumulating over the service life of the new asset, methods of dynamic investment calculus must be applied. There is a time lag and a cumulative effect of benefits. It should be noted that the benefits from rehab investments extend beyond a forecasting horizon of say 50 years according to the service life expectancy of rehabilitated assets, which are assumed to last more than 100 years on the average.

Benefits, which occur in the far future, do not have the same weight as costs that accrue in the next years. Therefore, costs and benefits are discounted into present values. As the present values of costs and benefits depend heavily on the discount rate, internal rates of return are calculated for which the discounted benefits are equal to the discounted costs. Due to the fact that rehabilitation benefits extend beyond the period of investment, there are significant net benefits at low discount rates. Rehabilitation investments pay off as long as future net benefits are discounted with rates beyond the internal rates of return. In general, it is recommended that the internal rate of return should be at least 3% larger than the inflation rate.

For this second option, the following output charts can be generated:

- Length or number of assets to rehabilitate, based on short, median and long service life expectancies of asset categories
- Annual rehabilitation rates for asset categories, based on short, median and long service life expectancies
- Annual rehabilitation rates for assets, based on short, median and long service life expectancies for asset categories
- Annual failure rates for do-nothing alternative and rehabilitation according to short, median and long service lives of asset categories
- Annual leakage rates for do-nothing alternative and rehabilitation according to short, median and long service lives of asset categories
- Development of the assets age and residual service life expectancy, based on short, median and long service life expectancies
- Annual repair costs for do-nothing alternative and rehabilitation according to short, median and long service lives of asset categories
- Annual water savings for do-nothing alternative and rehabilitation according to short, median and long service lives of asset categories
- Annual cost savings from reduced failures for different social cost factors
- Annual cost savings from reduced water leakage for different marginal costs

- Annual rehab investments, based on short, median and long service life expectancies
- Balance of investments and cost savings
- Internal rate of return for rehab investments, based on short, median and long service lives

The workflow model of the *Rehabilitation Strategy Manager* software is shown below.

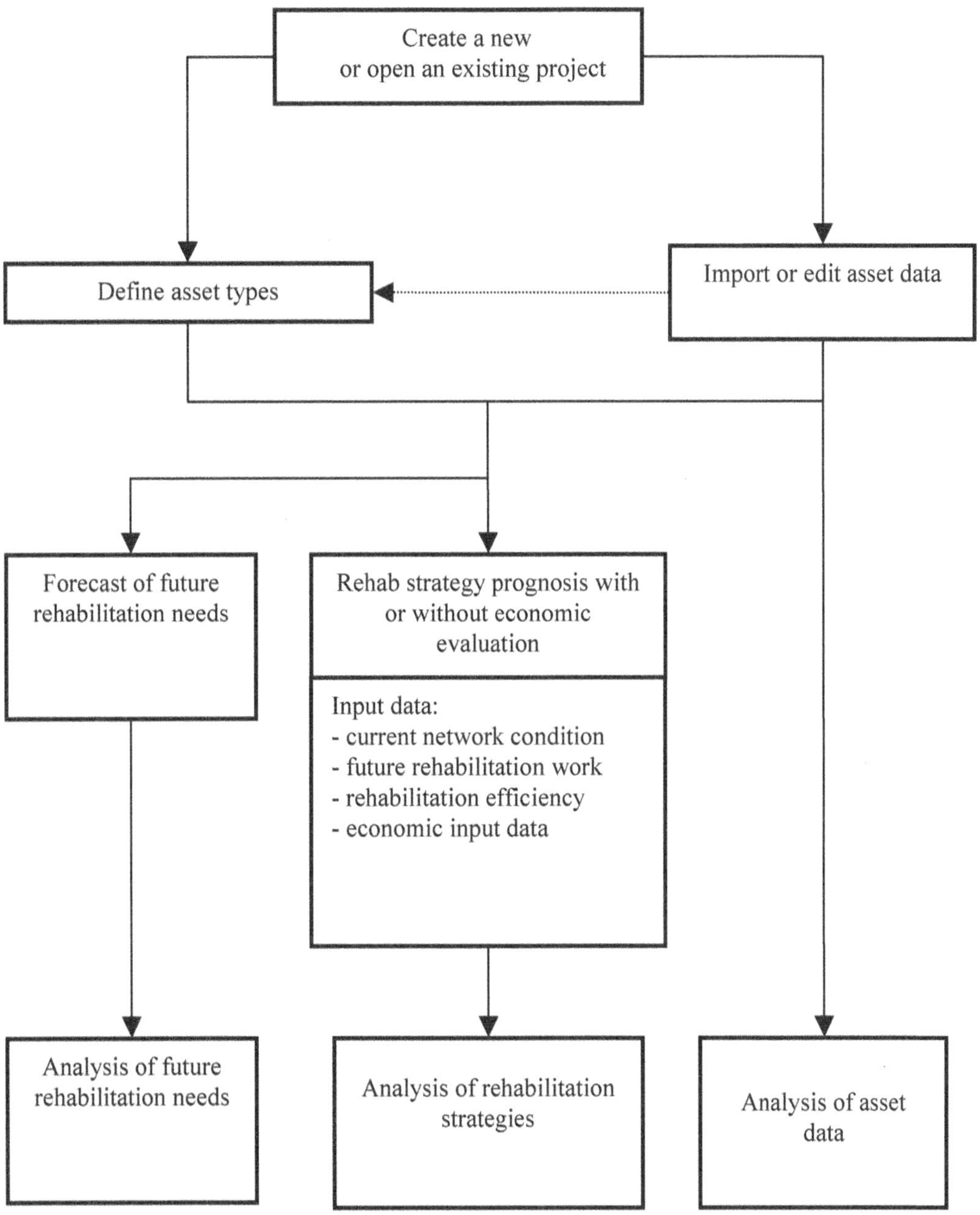

Figure 4.6. Workflow structure of the Rehab Strategy Manager software

4.7.1 Rehabilitation Strategy Evaluation

The Balancing and Ranking Procedure

The Balancing and Ranking procedure (BRP) is a new variant within the field of semi-formalised multi-criteria decision support methodologies. "Basic features of the approach are the pair wise comparisons of options, mixed scales and the so-called balancing principle, i.e. the balancing of vectors of advantages and disadvantages" (Strassert 2000). For any pair of options the relative advantages and disadvantages are balanced and, simultaneously, the different importance of the score is taken into account. The theoretical framework of the approach is briefly explained in the following.

Preliminary order of options

The procedure starts with the impact matrix (Figure 4.7). Each criterion C_i yields an individual order of options P_j by means of e_{ij}. For $e_{11} > e_{12} > e_{13}$, the ranking order of criterion C_1 is P_1, P_2, P_3. In general, the ranking order of options P_j will differ for all criteria C_i.

In an "outranking matrix" (Figure 4.7a), the number of pros and cons are counted for each option. The matrix in Figure 4.7a shows that option P_1 is placed twice (i.e. for two criteria) before option P_2, once before P_3, and vice versa P_2 is placed once before option P_1, and P_3 is placed twice before option P_1 (The table must be read by rows).

	P_1	P_2	P_3
P_1	0	2	1
P_2	1	0	1
P_3	2	2	0

Figure 4.7a. Outranking matrix

	P_3	P_1	P_2
P_3	*	2	2
P_1	1	*	2
P_2	1	1	*

Figure 4.7b. Triangular outranking matrix

When the "outranking matrix" is set up, the options must be re-ordered for achieving a triangular matrix (Figure 4.7b). Triangularisation is the systematic re-ordering of options such that out of a set of p = j! orders, the total sum of the values above the main diagonal is maximal. A situation where only zero values are below the main diagonal corresponds to a strong transitive overall final order of options, or total order structure. Normally, this total order structure of options does initially not exist.

The outranking matrix relates to the majority rule of counting votes (1 criterion → 1 vote). In the context of the balancing principle, the assumption of a majority rule must be given up. Instead, in the BRP the outranking matrix is used as a pre-order of options, which should be subject of a screening and balancing process.

4.7.2 The balancing process

Firstly, a new table is introduced. In an "advantages-disadvantages table" the criteria C_i are combined with the result of the comparison in pairs of options P_j/P_k (Figure 4.8). The head of the table contains all possible comparisons in pairs. For n options the number of comparisons is z = n*(n-1)/2.

	P_1/P_2	P_1/P_3	P_2/P_3
C_1	A_{P1C1}	A_{P1C1}	A_{P2C1}
C_2	D_{P1C2}	D_{P1C2}	A_{P2C2}
C_3	A_{P1C3}	D_{P1C3}	D_{P2C3}
ΣA_j	2	1	2
ΣD_j	1	2	1

Figure 4.8. Advantages-disadvantages table

For each comparison in pairs, the scores of e_{ij} must be compared. The comparisons can be made independently from the criteria's scales. They refer to quantities, rankings or frequencies (cardinal, ordinal, nominal scale). At the bottom of the table two rows show the sum of advantages ΣA_j and the sum of disadvantages ΣD_j by comparison.

In the advantages-disadvantages table, each column represents a separate binary decision problem: in the first column, the comparison P_1/P_2 of the two options P_1 and P_2, the question is whether the two advantages [A_{P1C1}, A_{P1C3}] together dominate (or not) the disadvantage D_{P1C2}. For "Yes", P_1 is strictly superior to P_2. If respectively the answer is "No", then P_2 is strictly superior to P_1. If the answer to the question: Are [A_{P1C1}, A_{P1C3}] strictly superior to $D_{P1C2}$is "Yes", the disadvantage D_{P1C2}looses its importance.

Thus, each comparison result is rated according to a binary code (1;0), and the results of all comparisons can be drawn to a compatibility matrix C (Figure 4.9a), and if the compatibility matrix is reordered, to the triangular compatibility matrix C^T (Figure 4.9b).

	P_1	P_2	P_3
P_1	0	1	1
P_2	0	0	0
P_3	0	1	0

Figure 4.9a. Compatibility matrix C

	P_1	P_3	P_2
P_1	0	1	1
P_3	0	0	1
P_2	0	0	0

Figure 4.9b: Triangular compatibility matrix C^T

The triangular compatibility matrix C^T is representing a "strict total order" with a binary relation, which is asymmetric, complete and transitive. With this matrix, the overall final order of options looked for is obtained, i.e. $<P_1, P_3, P_2>$.

4.7.3 Operating comparisons in pairs

In practice, the number of strategies to be compared can easily be around 5, 6 or more. Then the number of comparisons in pairs, z is 10, 15, or more, respectively. Irrespective the number of criteria (for the comparison of strategies in CARE-W the number of criteria is limited to 10), the determination of the order relations could become a clumsy job.

But not necessarily all possible comparisons in pairs and corresponding balancing problems must be carried out. For 4 strategies, the number of comparisons would be 10. The number of possible orders of options is p = n! = 1*2*3 = 24. These orders are:

P_1, P_2, P_3, P_4	P_2, P_1, P_3, P_4	P_3, P_1, P_2, P_4	P_4, P_1, P_2, P_3
P_1, P_2, P_4, P_3	P_2, P_1, P_4, P_3	P_3, P_1, P_4, P_2	P_4, P_1, P_3, P_2
P_1, P_3, P_2, P_4	P_2, P_3, P_1, P_4	P_3, P_2, P_1, P_4	P_4, P_2, P_1, P_1
P_1, P_3, P_4, P_2	P_2, P_3, P_4, P_1	P_3, P_2, P_4, P_1	P_4, P_2, P_3, P_3
P_1, P_4, P_2, P_3	P_2, P_4, P_1, P_3	P_3, P_4, P_1, P_2	P_4, P_3, P_1, P_2
P_1, P_4, P_3, P_2	P_2, P_4, P_3, P_1	P_3, P_4, P_2, P_1	P_4, P_3, P_2, P_1

If n-1 comparisons in pairs are already executed, and if all options are considered in the balancing process, the remaining comparisons are given implicitely. For example, if n - 1 = 3 comparisons yield the three orders $<P_1, P_2>$, $<P_1, P_4>$ and $<P_3, P_2>$, then from the 24 orders above, in a first step 12 are eliminated (all orders where P_1 is before P_2), in a second step 8, and in a third step 3 orders are eliminated, respectively. Hence, only one single order, $<P_1, P_4, P_3, P_2>$, remains that is the final order of options.

Intransitivities occur, when the balancing result of three comparisons would give rankings of the form: $<P_1, P_2>$, $<P_2, P_3>$ and $<P_3, P_1>$, i.e. the circular relation $P_1 > P_2 > P_3 > P_1$. Strassert gives an analytic proof that the number of comparisons in pairs could be reduced down to a minimum of n-1 comparisons, and a methodology how to avoid intransitivities in the final order. For the pre-elimination of options, a k.o. criterion could be introduced (e.g. maximum investment costs).

4.7.4 Solving balancing problems

The solution of the balancing problem is the comparison of two options with respect to the set of advantages against the set of disadvantages, presented in the "advantages-disadvantages table". That is the answer to the question: Are the advantages (or the bundle of advantages) $[A_{P1C1}, A_{P1C3}]$ strictly superior to the disadvantage (or a bundle of disadvantages) D_{P1C2}?

The number of criteria in each comparison with an equal score is not decisive. A dominance check looks for comparisons, where we have only advantages but no disadvantages. Rankings including the inverse of such a dominance order can be removed from the set of possible rankings.

4.8 PERFORMANCE INDICATORS FOR THE EVALUATION

According to the purpose of the *Rehab Strategy Evaluator* - the software to ease this task - the following decision criteria was chosen the indicators calculated for each rehabilitation strategy with the *Rehab Strategy Manager*:

- Network rehabilitation rate [%]
- Average age of network [years]
- Residual service life of network [years]
- Failure rate [no./km*year]

- Leakage rate [m³/km*h]
- Cost savings from reduced repair [€]
- Cost savings from reduced leakage [€]
- Investment costs [€]
- Year of Break Even [year]
- Internal rate of return [%]

The calculated results for the decision criteria 1-8 (see above) are included from the start year of the prognosis up to the end in 5-year steps. Finally the "year of break even" and the "internal rate of return" are fixed numbers valid for one strategy.

4.9 SOFTWARE: THE REHAB STRATEGY EVALUATOR

To ease this rather complicated multi-criteria evaluation process the *Rehab Strategy Evaluator* software was developed.

At first, all network rehabilitation strategy candidates are checked for knockout criteria, i.e. characteristics, which under no circumstances suit the future background scenarios of this particular utility. Probably the most important knockout criteria would be the rehab budget available, which depends on expected revenues and expenses and is strongly influenced by water consumption behaviour and population and industrial development.

Next step is the dominance test, checking whether one rehab strategy candidate is dominated in all relevant characteristics by another one. Dominated candidates are eliminated.

In general, each strategy has some advantages in some characteristics and disadvantages in other aspects when compared to another strategy in a particular scenario. These sets of advantages and disadvantages are compared in pairs, and balanced against each other; in order to decide which of the two candidates is to be preferred. A short example of such a table of advantages and disadvantages is shown below.

Table 4.5. Comparison in pairs – Balancing advantages and disadvantages of two strategies

Preferring the Low Budget Strategy (1) over the Long Durability Strategy (2) would have		
	Advantages (1) versus (2)	Disadvantages (1) versus (2)
Criteria	annual budget is 8 Mill € lower assets are used 20 years longer	failure rate is 0.3 higher leakage rate is 0.2 higher new assets live 40 years shorter
Result of balancing	Strategy (1) is preferred (in spite of its disadvantages) → (1) > (2)	

As the annual budget is one of the most important criteria, in this example, strategy 2 is outranked by strategy 1 in spite the fact that strategy 1 seems to have fewer advantages. However, even if costs and benefits were balanced for both strategies, in view of the economic key factor scenarios, it may be very uncertain whether the water utility will be able to spend so much on rehabilitation in the future. Thus, strategy 1 would be more robust. This may suffice to illustrate the case.

The last step of the procedure is to rank all strategies according to the results of pair wise comparisons. The number of comparisons is (n-1)n/2. Theoretically, there are 2 exp[n(n-1)/2] combinations of different results from the comparison of n alternative candidates. This may cause inconsistencies in putting together the final ranking order. Therefore, the *Rehab Strategy Evaluator* is checking for intransitivity and alarming the user if his preferences would not lead to a consistent final ranking order. Advice is given how intransitivity may be avoided. The workflow model of the software can be seen in Figure 4.9.

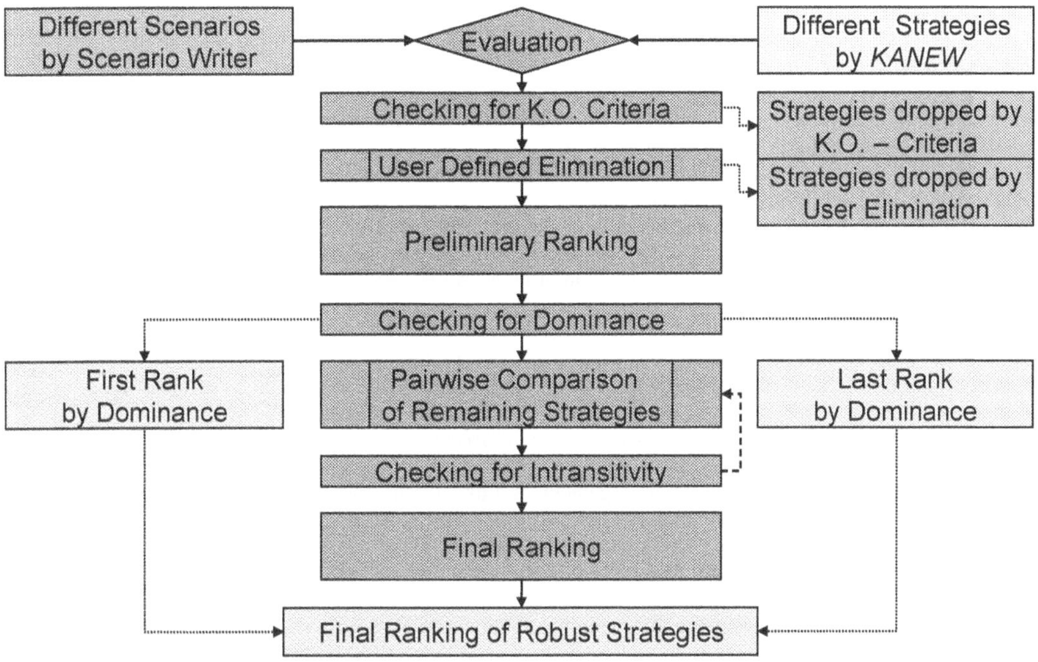

Figure 4.10. Workflow structure of the Rehab Strategy Evaluator software

4.10 PRACTICAL APPLICATION

The aim of work package 4 was to point out how the long-range effects of rehabilitation programmes in a comprehensive cost benefit framework can be used to develop an "optimal" network rehabilitation program. The process of network deterioration and the effects of rehabilitation needs that are not fulfilled in due time, are calculated with a cohort survival model. With this model, as implemented in the *Rehab Strategy Manager* software, several network performance indicators are forecast year-by-year and expressed in economic terms. Assumptions have to be made for such long-term forecasts and economic evaluation. The model output depends on specific assumptions on the assets' service lives and on the development of the water utility's economic key factors. Uncertainties of the future are taken into account by creating consistent scenarios for the utility background variables and by choosing a robust rehabilitation strategy.

The three software packages give decision support to create long-term scenarios (*Rehab Scenario Writer*), to explore the medium- and long-term effects of alternative rehabilitation programs or strategies (*Rehab Strategy Manager*) and to choose the best robust rehabilitation strategy for the future (*Rehab Strategy Evaluator*).

From previous research and case study applications, some general conclusions can be drawn. Long-term projections appear to be a useful complement to the "fire brigade" and monitoring approaches prevailing in water supply network management. In spite of limited forecasting capabilities the band of rehabilitation needs can be determined by a cohort survival model for specific asset types in the near and far future. Cohort survival models are not only implemented in the *Rehab Strategy Manager*, they are also at the core of *WARP* (Rajani/Kleiner 2001), the Nessie Model (Scharfenaker 2001) and in application for sewers. Such models should be used for asset management in water companies. Efforts are made all over the world to establish a framework for capital maintenance planning (AWWA 2001, Buckland 2000, UKWIR 2002).

The work package 4 approach is a top down approach, which is recommended for strategic investment planning and asset management. The focus is on the network level, not on individual assets within the system. The efficiency of a rehabilitation programme depends very much on the efficiency of the individual asset rehabilitation projects, which are part of the rehabilitation programme. So, in the top down approach, the next step is an important one: selecting the most cost-efficient rehabilitation projects. This research was done in working package 3 of the CARE-W project (Le Gauffre et al., 2002) and in a research project at TU Dresden (Baur/Kropp 2002; Baur, 2002).

4.11 REFERENCES

Alegre, H.; Hirner, W. Baptista, J.M.; Parena, R. (2000): Performance indicators for water supply services, Manual of Best Practice Series, IWA Publishing, London.

Alegre, H. (2002): Performance Indicators for Network Rehabilitation. In: International CARE-W Conference on Computer Aided Rehabilitation of Water Networks, pp.53-64. Dresden.

AWWA (2001): Dawn of the Replacement Era. Reinvesting in Drinking Water Infrastucture. Denver.

Baur, R. (2002): Bestimmung kosteneffizienter Rehabilitationsprojekte – dargestellt am Beispiel städtischer Wasserrohrnetze (Selecting cost-efficient rehabilitation projects – case of urban drinking water networks). PhD Thesis, Department of Civil Engineering at TU Dresden.

Baur, R. and I. Kropp (2002): Computer Aided Rehab Budget Allocation - A Case Study from Germany. Proceedings of IWA Specialized Conference "Management of Productivity at Water Utilities". Praha.

Baur, R. and R. Herz Eds. (1999): Service Life Management of Water Mains and Sewers. Proceedings of 13th European Junior Scientist Workshop, Dresden.

Baur, R., I. Kropp and R. Herz Eds. (2003): Rehabilitation management of urban infrastructure networks. Proceedings of the 17th European Junior Scientist Workshop, Dresden.

Buckland, P. (2000): Risk based economic life cycle management of infrastructure assets and its regulatory implications. Proceedings of International Conference of Maintenance Societies, Wollongong, Australia.

CARE-W WP4 – Strategic Planning and Investment (2002): Reports D9, D10 and D11.

Deb, A., Y.J. Hasit, F. Grablutz and R. Herz (1998): Quantifying future rehabilitation and replacement needs of water mains, AWWA Research Foundation, Denver.

DVGW Regelwerk (1997): W401:Entscheidungshilfen für die Rehabilitation von Wasserrohrnetzen (Decision support for the rehabilitation of water distribution systems). 26 p.

DVGW Regelwerk (1999): G401:Entscheidungshilfen für die Rehabilitation von Gasverteilungsnetzen (Decision support for the rehabilitation of gas distribution systems). 29 p.

Herz, R. (1994): Alterung und Erneuerung von Infrastrukturbeständen – ein Kohortenüberlebensmodell (Ageing and rehabilitation of infrastructure assets – a cohort survival model). Jahrbuch für Regionalwissenschaft 14, pp. 9-28.

Herz, R. (1996): Dégradation et renouvellement des infrastructures – un modèle de survie par cohorte. Flux 23, pp. 21-36.

Herz, R. (1996): Ageing processes and rehabilitation needs of drinking water distribution networks. J Water SRT - Aqua 45, 5, pp. 221-231.

Herz, R. (1998): Exploring rehabilitation needs and strategies for drinking water distribution networks. J Water SRT - Aqua 47, 6, pp. 275-283.

Herz, R. and A. Lipkow (2002): Strategic Water Network Rehabilitation Planning. CD-ROM Conference Proceedings of 3rd IWA World Water Congress, Melbourne.

Herz, R. (2002): Software for strategic network rehabilitation and investment planning. In: International CARE-W Conference on Computer Aided Rehabilitation of Water Networks, pp.65-84. Dresden.

KANEW: www.tu-dresden.de/biwiss/stadtbau/KANEW.html

Kleiner, Y. and B.B. Rajani (1999): Using limited data to assess future pipe replacement needs. J AWWA, 91(7), pp. 47-61.

Le Gauffre, P., R. Baur and K. Laffréchine (2002): Multicriteria decision support for annual rehabilitation programmes. In: International CARE-W Conference on Computer Aided Rehabilitation of Water Networks, pp.85-95. Dresden.

Lipkow, A. and R. Herz (2002): Scenario Based Evaluation of Long-term Rehabilitation Strategies. Proceedings of IWA Specialised Conference on "Management of Productivity at Water Utilities", Praha, pp.139-144.

Rajani, B.B. and Y. Kleiner (2001): 'WARP - Water Mains Renewal Planner'. In: International Conference on Under-ground Infrastructure Research, Waterloo, Canada.

Rajani, B.B. and Y. Kleiner (2002): Towards Pro-active Rehabilitation Planning of Water Supply Systems. In: International CARE-W Conference on Computer Aided Rehabilitation of Water Networks, pp. 29-38. Dresden.

Scharfenaker, M.A. (2001): Mythic monster aids first scientific assessment of nationwide infrastructure needs. AWWA J., July 2001 pp. 20-32.

Strassert, G. (2000): *The balancing principle, strict superiority relations, and a transitive overall final order of options.* Diskussionspapier Nr.34, Institut für Regionalwissenschaft der Universität Karlsruhe. 19p.

UKWIR Ltd (2002): A Common Framework for Capital Maintenance Planning in the UK Water Industry.

5

Development of a water network rehabilitation manager

Paul Conroy, Joanne Hulance, Mark Kowalski and Kevin Taylor, WRc

5.1 BACKGROUND

The objective of the CARE-W project is to establish a clear and logical framework for water network rehabilitation decision-making. The ultimate product is a set of methods and software, enabling rehabilitation decision-makers to establish and maintain effective management of their water networks; in other words, to *rehabilitate the right pipelines at the right time.*

CARE-W is a methodology for rehabilitation planning, which comprises a written procedure, a number of software tools, and a Water Network Rehabilitation Manager that provides a means of integrating the various software tools through common data management and reporting protocols. The CARE-W project comprises a suite of powerful network management tools, designed to identify failure patterns and cost-effective rehabilitation requirements for water distribution networks. A more powerful and comprehensive approach to rehabilitation planning is possible through the harmonisation and optimisation of these tools.

Work Packages (WP) 1 to 4 of the CARE-W project have focused on the identification, design and development of this suite of tools, aimed at modelling and predicting typical problems such as structural failures, hydraulic insufficiencies, leakage, deteriorating water quality and increasing maintenance costs. WP5, however, integrated these tools into the CARE-W Water Network Rehabilitation Manager.

The specific objectives of WP5 are as follows:

- Develop the CARE-W procedure for optimising rehabilitation planning;
- Define the structure of the Water Network Rehabilitation Manager Software, and;
- Create the Rehabilitation Manager software.

To develop the CARE-W procedure the individual CARE-W tools have to be thoroughly evaluated in terms of their benefits, data requirements and inter-relationships. In some instances there is some overlap of functionality and differences of approach under specific local conditions. The benefits of the application of these tools must be determined and a procedure for optimum integration developed. The procedure must clarify how to utilise the CARE-W toolbox most efficiently in order to address a given problem or circumstance.

The key performance indicators and data flows must be developed and specified as a prerequisite to development of the software. The structure of the Rehabilitation Manager software will depend upon the data outputs and reports that are required and output by the CARE-W tools. It is therefore essential that this information be thoroughly specified.

The Rehabilitation Manager software must support the integration of and data flows between the CARE-W component tools. The software will allow automatic and manual data input and store information (results of analysis) that can, for example, enable write back to electronic corporate data storage systems. A manual has also been produced to support the use of the software.

5.2 SCIENTIFIC METHODOLOGY APPLIED

The starting point for this work package was to understand the differences and synergies between the tools that comprise the CARE-W tool-kit. The nature and benefits of each needed to be fully understood in order to evolve a procedure for development and optimal, integrated application of the tools. Once the theoretical procedure for operation had been specified, the task of integrating the software and modeling tools could begin. Data input, output, storage and handling were specified.

The next stage was to design and specify the Rehabilitation Manager software such that the necessary data flows could be accommodated. It was important to structure and configure this software so that it provides an effective conduit between corporate data systems and the CARE-W suite of tools. Finally, the software was created, tested and delivered in a fully functional state of operation.

The more detailed methodology adopted to deliver the CARE-W procedure and Rehabilitation Manager software is outlined below.

5.2.1 The CARE-W Procedure

When the CARE-W project started, there were already a number of pre-existing tools and approaches. These varied in levels of sophistication and functionality according to local market requirements. Consequently, at an early stage in CARE-W's development, it was necessary to develop a general framework for rehabilitation planning that was flexible enough to address a range of European needs, whilst still allowing the user to focus on local or national requirements. Major issues included the variability in data quality and availability, and differences in planning objectives. CARE-W must be viable across Europe and not just for the minority of water service providers who hold good-quality data. Furthermore, the tools must offer benefits for small municipalities as well as the large, private water companies.

This generic rehabilitation procedure is summarised in Figure 5.1 below.

Detailed assessment of the methodology, objectives and outputs from the CARE-W tools led to the development of the specific CARE-W procedure. This procedure acknowledges all of the recognised approaches and variants on the rehabilitation planning process. It encompasses strategic, tactical and budget-limited methods for rehabilitation planning. The flexibility of CARE-W is such that new methodologies can be incorporated, should they be specified by the end-user.

As well as strategic and tactical tools, the need to invest effectively up to a budget ceiling or the need to define the budget have been identified. This reflects different end-user expectations and is achieved by using specific tools in a particular sequence.

Sequential use of the tools requires the CARE-W toolkit to be an integrated system. One of the special features of the toolkit is that decision-making information can be passed from one tool to the next via the Rehabilitation Manager. This was a critical step in achieving integration and tool interaction.

The CARE-W procedure provides an effective means of using the CARE-W tools. The tools can be used individually or in various combinations at the behest of the user – it is this flexibility that makes CARE-W such an important development.

Details of the development of the CARE-W Procedure are presented in CARE-W report D12: The CARE-W Procedure, published in October 2002.

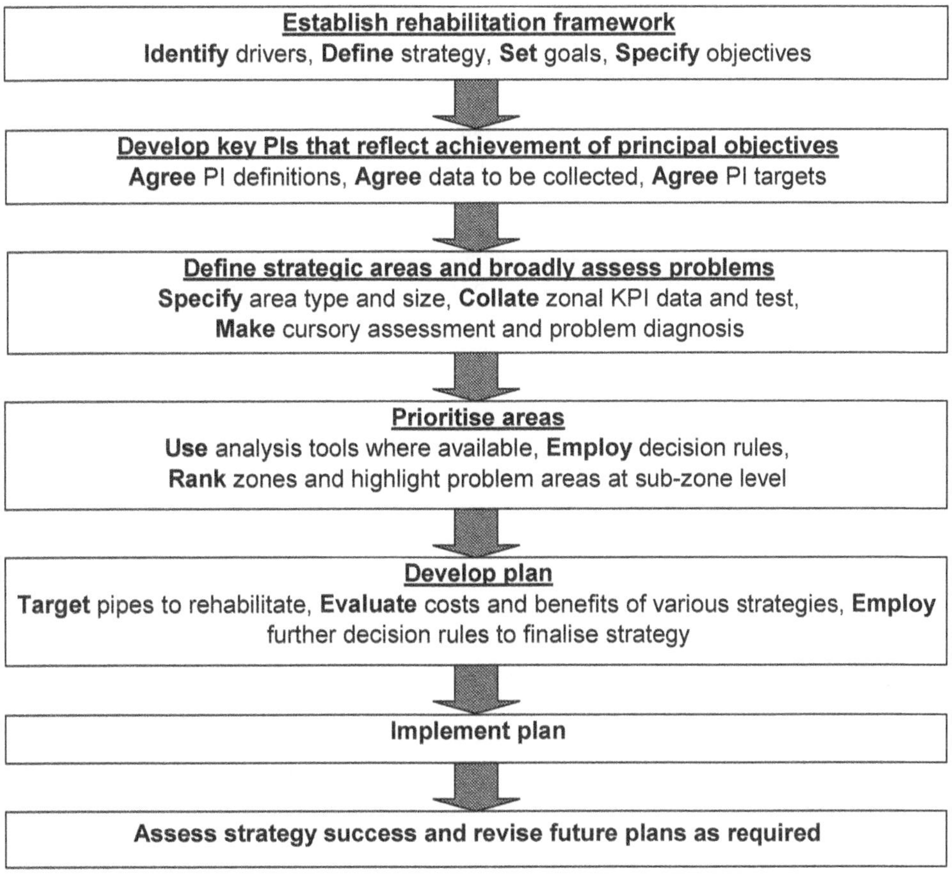

Figure 5.1. Generic Rehabilitation Planning Procedure

5.2.2 The Water Network Rehabilitation Manager Software

A number of specific steps were taken in the development of the Rehabilitation Manager Software, as follows:

- Specification of the data input/output and storage requirements for the integrated CARE-W package;
- Production of a specification for the software that will integrate the use of the tools defined in CARE-W in an optimal manner;

The objective was to deliver working software, which will manage the information needed and results generated from the tools defined in CARE-W thus enabling the integrated CARE-W package to identify optimum rehabilitation strategies.

The Rehabilitation Manager software is a versatile and powerful application, incorporating a variety of rehabilitation planning tools. It has the flexibility to allow the rehabilitation of networks to be planned at both strategic and tactical levels, or in other words, at various levels of complexity: network, zone or pipe. This flexibility is possible because of the nature of the tools that form the software.

Details of the development of the Rehabilitation Manager software are presented in CARE-W report D13: The Architecture of the CARE-W Rehabilitation Manager Software.

5.3 SCIENTIFIC RESULTS OBTAINED

There were two specific outputs from WP5, the CARE-W Procedure for Rehabilitation Planning and the Water Network Rehabilitation Manager Software. These two outputs are described in more detail below.

5.3.1 The CARE-W Procedure

Efficient planning requires the rehabilitation engineer or planner to be in possession of a substantial amount of background knowledge and experience on the types of problems faced, current performance and possible effective solutions. The engineer must be aware of the objectives of rehabilitation for each problem he faces and apply sound judgments using all the tools at his disposal in an appropriate manner. This places a huge burden on the engineer when many rehabilitation methods are feasible and there are many solutions to improve service delivery to customers. The engineer can narrow his search for an acceptable solution by using strategic level data and knowledge of historical performance. However, scheduling rehabilitation over, say, a 2-year planning horizon is a complex multivariate problem and a suite of useful analysis tools to help him choose the most cost-effective rehabilitation options from those available must guide the engineer.

There are a number of options for developing rehabilitation plans and these are reflected in the CARE-W Rehabilitation planning procedure. The Rehabilitation Manager software will be of greatest use when some initial analysis of the problems identified in the target network has been carried out. This groundwork is necessary in all but the most cursory of planning exercises. The options discussed below therefore assume that the Engineer has completed the initial development stages of his rehabilitation planning exercise, i.e. defining and prioritising the areas of the network that require further evaluation.

There are at least three main alternative routes when undertaking a rehabilitation planning exercise. Different options will be valid in different circumstances and are likely to depend on the data available:

Option 1: All planning at strategic level

The rehabilitation strategy is derived from the *top down*; global spending levels are set strategically and broad areas for more detailed analysis are identified. A strategic rehabilitation plan is created with, possibly, a number of alternative rehabilitation scenarios.

Option 2: Strategic budgets set and pipe level targeting up to budget

The budgets for the rehabilitation schemes are set at a global level, i.e. for the utility or region. Rehabilitation schemes are then developed for particular areas or zones using more detailed knowledge and information. Schemes are only planned up to the ceiling set by the strategic budget. As investment is constrained by the strategic budget, the most critical problems will be given priority and further information may be sought to finalise those schemes that address these problems. Some negotiation is likely before plans are finalised.

Option 3: Pipe level build up of costs

A number of likely rehabilitation schemes are planned based on detailed pipe-level information for a given zone or area, and the costs and benefits summed over all pipes [*bottom up analysis*]. Likely schemes are prioritised and the rehabilitation plan for the year is negotiated.

Options 1 to 3 are now explained in the context of the CARE-W toolkit. They are also shown in colour-coded form in Figure 5.2.

Option 1: Strategic Rehabilitation Planning

The main element of the Rehabilitation Manager utilised by this first rehabilitation option would be the long-term rehabilitation planning tool developed under CARE-W Work Package 4 (WP4). This planning tool will allow engineers or planners to assess the effects of particular rehabilitation strategies using the cohort survival model for annual network forecasting.

The WP4 tool consists of three elements; the Scenario Writer, the Rehabilitation Strategy Manager and the Rehabilitation Strategy Evaluator.

The Scenario Writer assists the engineer or planner to write scenarios for the future condition and performance of a water supply system and the assets contained within it. These scenarios describe the interaction of various factors that will influence the behaviour of a network, over a 10-25 year planning horizon. The Scenario Writer creates three paths for the future of a given network, the best-case scenario, the worst case and the most likely scenario. Each of these paths will produce information relevant to rehabilitation planning and the key factors that will influence the rehabilitation policy.

The Rehabilitation Strategy Manager will allow the long-term effects of rehabilitation policies or programmes to be simulated, in terms of changes to specific performance indicators. The Rehabilitation Strategy Manager would be applied to a network as a whole, or to specific areas or zones. That network can consist of various types of asset, such as pipes of various materials, valves, service connections, etc. The Rehabilitation Strategy Manager is based on hard data about the network and the rehabilitation policies that can be applied. The results would be expressed as a cost for a particular rehabilitation programme at a given point in time.

The Rehabilitation Strategy Evaluator will compare the various rehabilitation options developed by the Rehabilitation Strategy Manager based on the scenarios created by the engineer or planner using The Scenario Writer. The Evaluator will incorporate a multi-criterion decision- making function which will allow the selection of the most appropriate rehabilitation strateg(ies).

A key output from the WP4 suite of software tools will be strategic budgets for rehabilitation scenarios applied to a pre-determined planning horizon and dataset, typically at network or sector levels.

Option 2: Annual Rehabilitation Planning Within Strategic Budgets

As stated earlier, the objective of this planning option is to use strategic modeling (via the Strategy Manager of WP4) to set global budgets. Detailed planning is then carried out to determine annual rehabilitation needs for specific zones, whilst maintaining the 'hard' constraint of the strategic budget figure.

The first part of this procedure would therefore follow that set out in Option 1 above, to determine the strategic spend profile.

Once this information has been determined the WP1 tool can be used fully to re-calculate the PIs for a particular zone. In this instance the WP1 tool would be used at its most detailed level where data allows. Typically this would be a cluster of a few pipes, but more commonly, a small sector of the network, such as a District Meter Area.

Where geospatial information is available, the GIS viewer incorporated into the Rehabilitation Manager software could be used at this point to view the elements of a particular zone that would be of interest to the rehabilitation planner. PI results for the zones of interest could be displayed. Alternative selections of sub-zones could be made and a new refined dataset created such that the working dataset contains only those pipes for further analysis. Standard methods of finding and selecting geospatial entities can be used, allowing the planner to select specific elements of a zone to incorporate them into the overall rehabilitation scheme.

The user has the option to include failure forecasting and hydraulic reliability data obtained from the WP2 tools at this stage. If the user has the data available, he may also use rehabilitation direct cost and benefit analysis results to supply the annual rehabilitation planning (ARP) tool with another criterion to judge which pipe categories to rehabilitate.

With or without WP2 and/or cost benefit analysis (CBA), the user may then initiate the WP3 ARP tool which will take the definition of the dataset (a network, sector or cluster of pipe identifiers), together with associated performance, utility and context information, to rank pipes in order of rehabilitation need, i.e. the priority of action (high, medium or low priority). If CBA is used, additional information could be provided on when assets need to be rehabilitated, and the rehabilitation method that should be employed.

The user must ensure that the WP3 tools will meet the specific objectives set out by the rehabilitation management framework. It is important that the user's key drivers are included in any decision support matrix. Only then can the ARP tool in WP3 rank the pipes according to the requirements set out in the original framework.

At this stage the engineer would review the data before finalising the rehabilitation plan, possibly by comparing outputs from different tools and a number of model runs with slightly different assumptions and boundary conditions. The final rehabilitation strategy will therefore be determined using engineering knowledge and experience of the operation and performance of a particular zone.

Option 3: Detailed Zonal Rehabilitation Planning

This option is a refinement of Option 2 and focuses on the detailed rehabilitation planning of specific zones. One key difference is the approach to prioritising investment. Budgets are determined from the rehabilitation requirements of each zone (bottom-up). There is no initial budget capping using strategic information.

Additional information on failures, maintenance and complaints data, plus hydraulic information would be sought. These would be used to supply failure forecasting and hydraulic reliability models developed as part of WP2. These tools produce a prediction of failure rate in years to come, or a hydraulic reliability based on information gleaned from hydraulic model data.

As well as the PI results applicable to the working dataset (refer to Option 2), WP2 modelling results may be combined with additional information from the user's experience (their *knowledge base*) plus a range of optional contextual *external information* (EI) to supply the Annual Rehab Planning MCD tool in WP3 with a wealth of detailed pipe-based information. Again, the ARP will produce a prioritised ranked list of pipes according to the points of view chosen as user preferences.

Finally, the detail of the ARP outputs may be combined with the results of a strategy-based cost-benefit analysis to allow the mix of rehabilitation schemes to be negotiated. The costs and benefits of the chosen schemes may then be summed and the rehabilitation and investment plans devised.

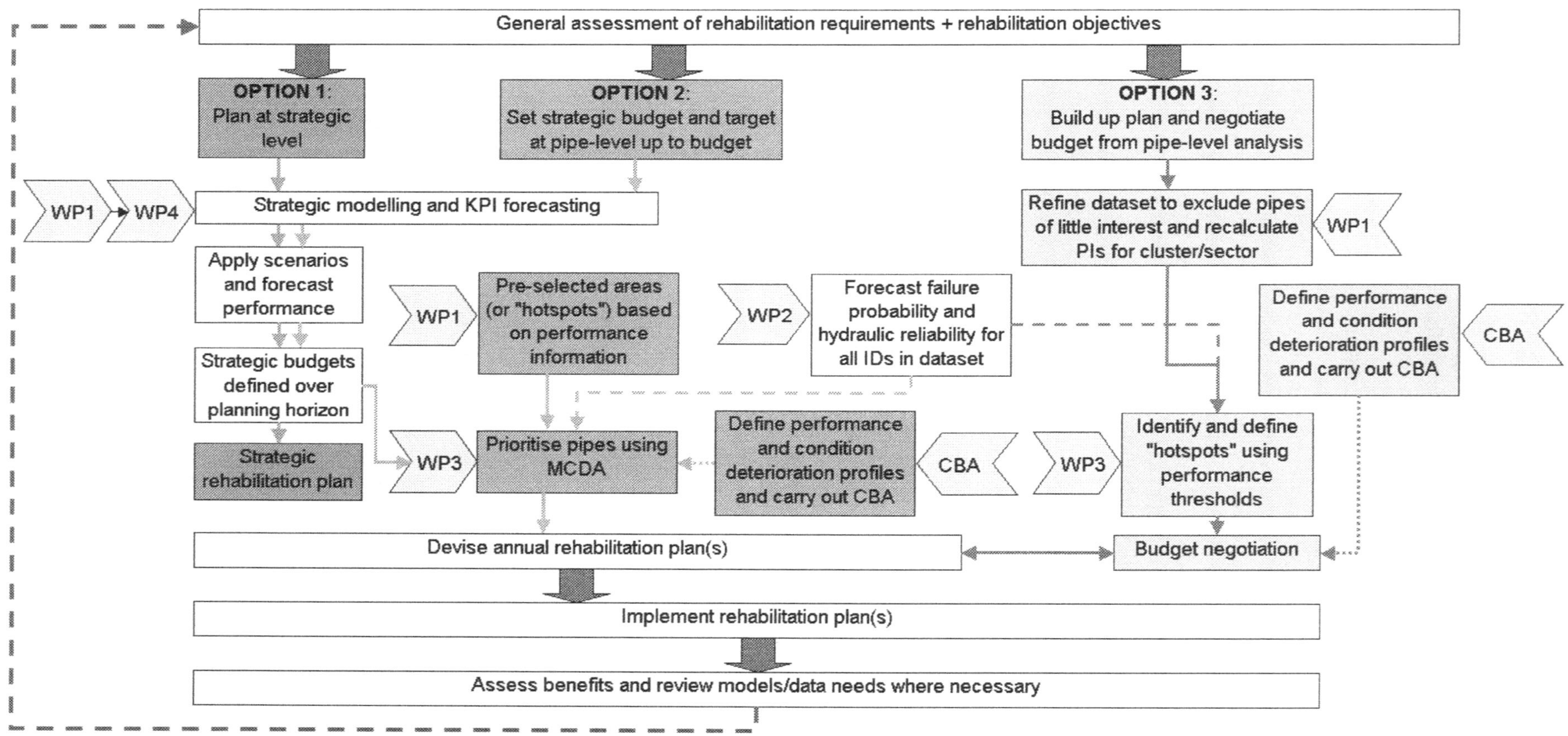

Figure 5.2. Rehabilitation Planning with CARE-W Tools

5.3.4 The Water Network Rehabilitation Manager Software

At the core of the CARE-W project is the Rehabilitation Manager. The Rehabilitation Manager software consists of a central MS Access 2000 database Visual Basic 6.0 application, which has the following attributes:

- Provides a central storage area and reference point for essential CARE-W data;
- Accepts **user-validated** input data in a pre-defined format. Data source fields will be mapped to fields within the CARE-W database according to the specification of the declared input data source;
- Provides automatic conversion of data item units to standard CARE-W convention;
- Performs rudimentary data validation including duplication, data type and range checking;
- Allows user interaction with a pre-determined range of tools under the CARE-W umbrella;
- Interacts with a pre-determined range of tools at the user's request by:
  - o creating the necessary input file(s) from the data stored in the CARE-W database to enable the tool to be run, and
  - o accepting and storing necessary output data on completion of tool use.

The general architecture of the Rehabilitation Manager software is shown in Figure 5.3.

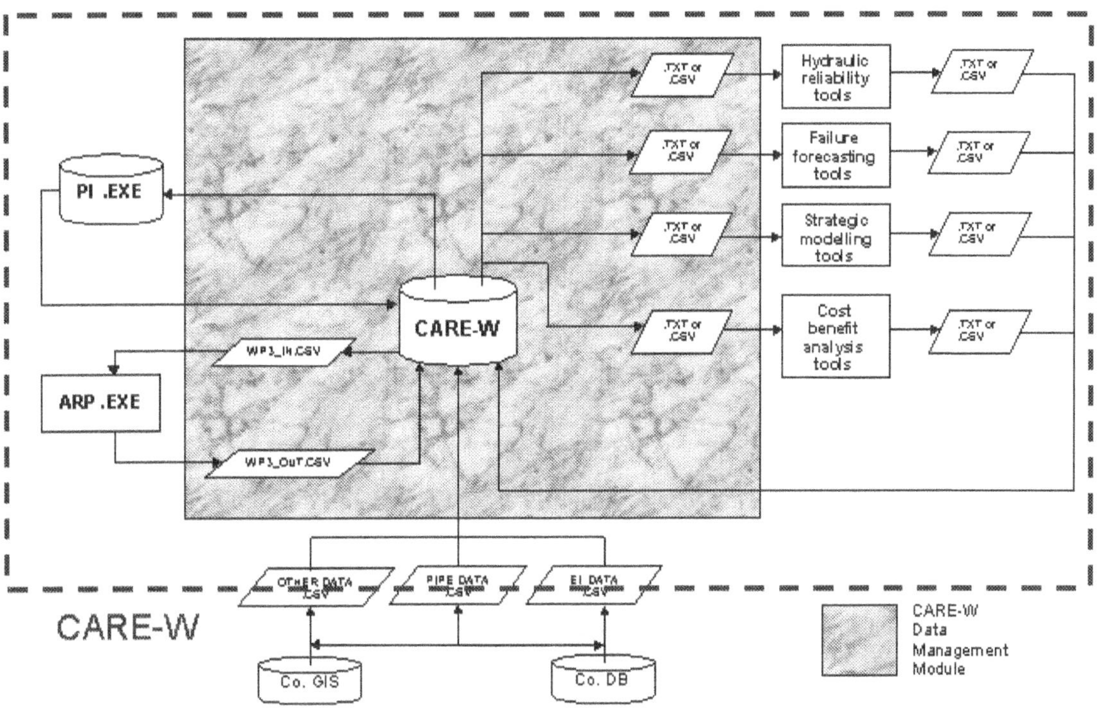

Figure 5.3. Rehabilitation Manager Architectural Design

The Rehabilitation Manager software has three key elements:

1. The ability to store key data in a managed central repository
2. The ability for various rehabilitation planning tools to interact and share data
3. The ability to use a number of rehabilitation planning tools in an integrated way

The Central CARE-W Database

The heart of the CARE-W software is its central database. This database consists of a series of tables that store specific types of information, such as pipe-based data, zonal, geospatial or hydraulic information, and results. The necessary link between input and results data is maintained to allow proper reporting of analysis history and results to the user.

Each database table contains a range of fields that relate to the various parameters or data fields used by each tool. Many of these fields, such as pipe diameter, length, and material, are used by a number of different tools. Others are unique to a specific tool. CARE-W has adopted its own standardised set of parameters and properties in order that the software can communicate with all analysis tools.

At the request of the user, the CARE-W software creates the necessary input file(s) for an analysis tool according to a pre-defined specification held internally. These files may then be used to run the analysis tool within (or outside) the CARE-W framework. Similarly, the software enables analysis results and any modified input data to be transferred back to the CARE-W central database and stored there for future reporting.

Data flow and tools interaction

The CARE-W database is a cornerstone in the integration of a number of tools, which in turn provides users with alternative routes to prepare their rehabilitation strategies and plans for the years ahead. This is not possible, however, without a facility to move data between the central database and the tools and vice versa. The main benefit of the CARE-W software to the rehabilitation planner is the power and flexibility of data handling and analysis. *The software must allow the user the option of using different tools to ensure a complete analysis and preparation of their rehabilitation plan.*

The software has a number of facilities for storing and processing data, some details of which are discussed below. However, it must be said that any analysis undertaken using the CARE-W software is only ever as good as the quality of the input data used in that analysis. It remains the user's responsibility to ensure that valid data are available for import, that is, the data are arranged in a consistent and definable manner, data types are correct, and data values are meaningful.

That said CARE-W does provide some rudimentary data validation. For instance, the fields within the central database tables have set units (m, mm, MPa, etc.) and data types (text, integer, binary [yes/no], etc.). On import to the database tables, the data are automatically converted to the standard units required by the software. Conversely, on export the data are converted to the units required by the individual CARE-W rehabilitation planning tools (although it is noted that, in most cases, the units required by the software are consistent with those required by the tools). The software will also check the data type as it is imported and check that it matches that of the recipient field in the CARE-W database. The user will be notified of any data format violations during validation.

It is envisaged that later versions of the software will undertake further basic data validation checks on data as they are imported into the central database. The data could be checked to ensure that key parameter values fall within reasonable tolerances. For instance, if it were known that pipe diameters in an area do not exceed 900mm and pipes are found in the dataset whose diameters are reportedly greater than 900mm, the user would be prompted to check the validity of the data presented.

Data may be imported to CARE-W from three sources, namely:

- CSV data files, e.g. mains database or maintenance records;
- Pre-existing analysis tool data files, e.g. 'SDF/MDF' failure forecasting files;
- GIS data files, e.g. ArcView 'shape' files.

Data may be imported into a project database in any number of steps and at any time. A mixture of data sources is also possible, and indeed likely, particularly where data have not been digitised or do not relate to a geospatial reference frame.

Most interaction with the range of tools included in the CARE-W software is managed via the Tools menu. At the request of the user, the CARE-W software creates the necessary input file(s) for an analysis tool from data held in the project database according to a pre-defined specification (managed by the internal Specification Manager). These files may then be used to run the analysis tool within (or outside) the CARE-W framework. Similarly, the software will enable the results of the analysis to be exported back to CARE-W and stored as necessary on completion of the analysis task.

The tools on the Tools menu are grouped according to their analysis function, namely: failure prediction, hydraulic reliability, multi-criterion decision support, cost benefit analysis and strategic planning. The specific tools are listed under these functions. For each tool there are a number of generic functions, namely:

- accessing tool-specific help;
- generating of input files from data in the project database (in order that the tool may be used to analyse the active dataset);
- editing the newly created data files for use by the tool;
- running the tool;
- importing the analysis results back into the central database;

Once the user has selected a tool to use for his analysis, data files must be created which will allow the necessary data to be used by the tool. The user may decide which optional data to include in the input files for a particular tool or use one of a number of pre-defined specifications. This specification will include the automatic conversion of units to those required by the individual tools, if different to those in the central database.

The edit facility allows the user to view and edit the import files created by the software before they are used by the rehabilitation planning tools. This allows the user to check the integrity of their contents and to verify successful file creation. CARE-W uses the system editor or MS Excel for editing as appropriate. To ensure consistency between input data and results files, if the input data files are edited interactively by a rehabilitation planning tool, the edited input data will be returned to the CARE-W database with any analysis results (output) data.

- The software is designed to accept and store results from analysis using CARE-W tools, wherever possible. The user is informed of the result of every data transfer process such as returning failure results from the PHM tool (Figure 5.4). The results returned are values required for use by other CARE-W tools and/or values which must be reported to the user directly, for example:
- *predicted failure and/or burst rates* calculated by the failure prediction tools
- *hydraulic criticality indices* calculated by the hydraulic reliability tools
- *Performance Indicator* results from the PI tool
- pipe rehabilitation priority *status* calculated by the ARP tool
- medium to long term *budgets* from strategic rehabilitation planning

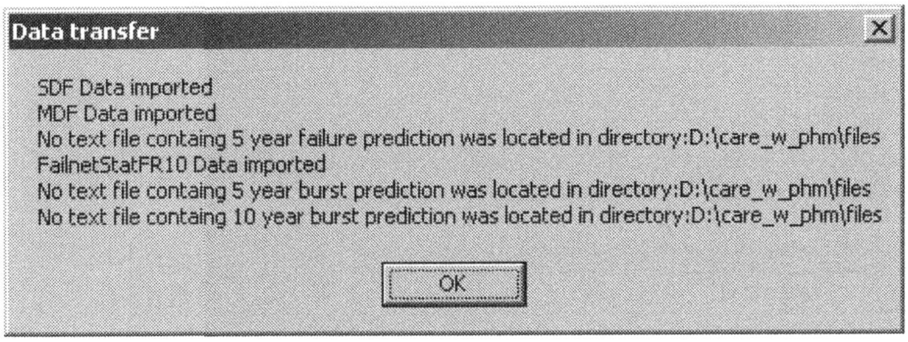

Figure 5.4. Example of data transfer report to user

Using the Rehabilitation Manager software

As stated previously, CARE-W supports a number of input formats including hydraulic model data in EPANET format, shape format GIS files and files created using the tools that are packaged with the CARE-W toolkit. These import types will be extended if the commercial need arises. Figure 5.5 illustrates the ease with which data of various pre-defined formats may be imported to a CARE-W project.

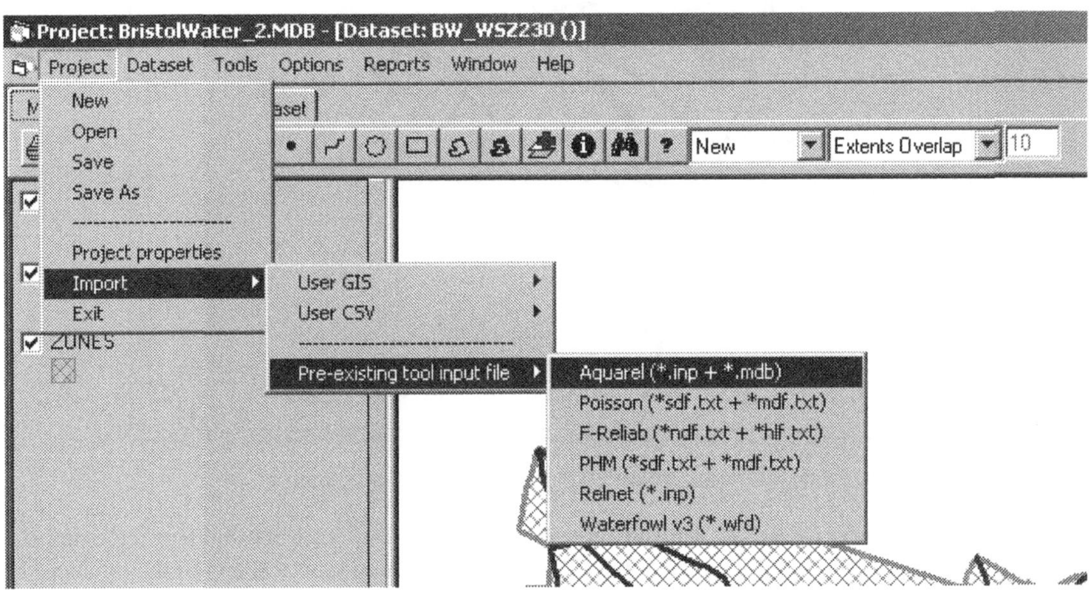

Figure 5.5. Importing data of various formats into a CARE-W project

The data contained within a project can be viewed in one of two ways, either geospatially by means of the in-built GIS viewer, or in tabular form. The geospatial view is the default view and becomes active once positional data have been imported to a CARE-W project.

The GIS viewer provides a pictorial representation of the data contained within a project, an example of the power of visualisation is demonstrated by Figure 5.6. As in standard GIS applications, the display is constructed from a series of layers which may be toggled on and off by checking the appropriate check box. In CARE-W, the standard layers are *All Pipes* (all pipes in the project are shown), *Labels* (pipe labels toggle), *Failures* (failure data toggle), and *Active Dataset* (current dataset highlighted).

In addition, there is the concept of an *active layer*. All actions (e.g. selection) are carried out on the active layer. This is the layer that is highlighted in the left-hand pane of the GIS viewer (the 'All pipes' layer in Figure 5.6). There are also layers for results data and dataset boundaries, where applicable.

A standard colour scheme has been adopted which is editable by the expert user. Pipes are shown in **mid-blue**, unless they are part of a user selection when they turn **khaki**. The constituent elements of the active dataset are shown in **purple**. Sector dataset boundaries are shown in **brown**.

The GIS viewer provides the user with the following functions:

- Zoom in/zoom out of specific geographical areas;
- Move the view to various geographical areas (pan);
- Select individual assets or groups of assets;
- Add/remove labels for assets;
- Create/remove dataset boundaries;
- Add/remove background maps.

GIS functions are accessed via the GIS toolbar above the map window. Most of these functions are generic to geographical information systems and thus will not be explained further here.

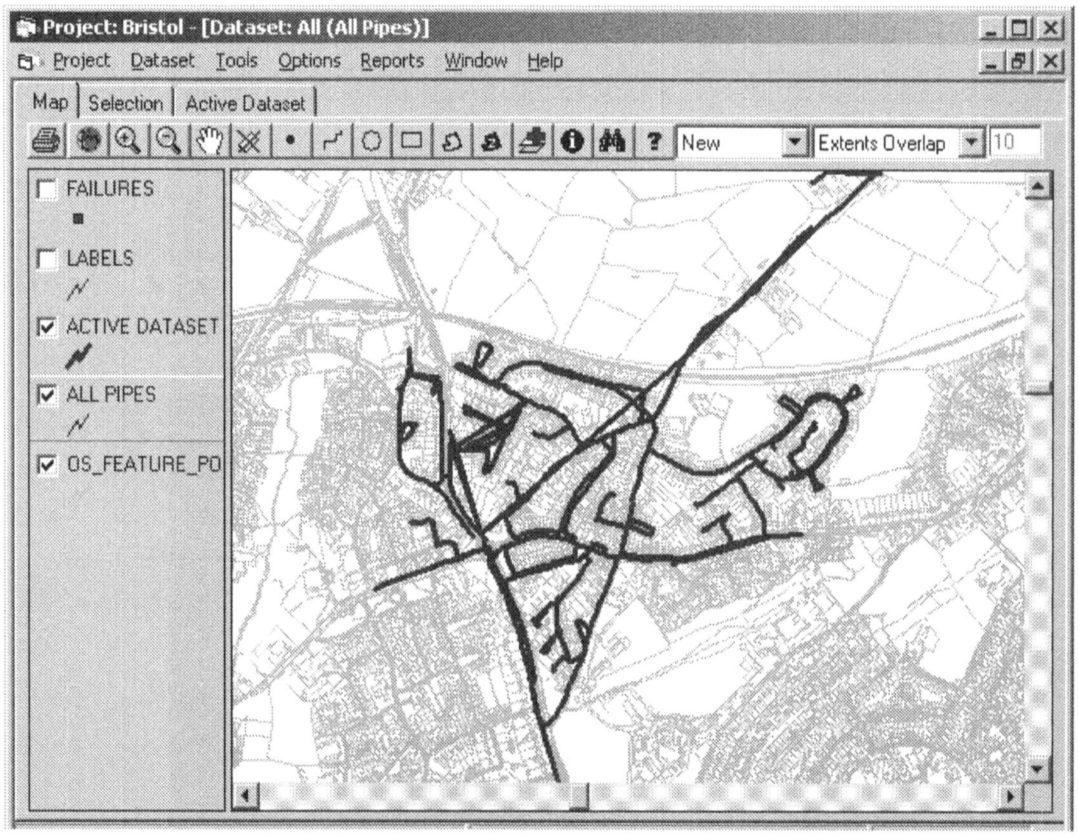

Figure 5.6. GIS viewer showing active dataset (purple-coloured pipes) and background map

Using the various tools within the CARE-W software is very straightforward. Once the links to the external rehabilitation planning applications are established using the CARE-W *Tool Manager* option, any of the referenced tools may be used with the Rehabilitation Manager. A dedicated Tools menu has been developed which allows the user to carry out the principal functions of data preparation, editing and retrieving results with ease (Figure 5.7). Once the analysis tool has been used, CARE-W retrieves any interactive changes in the input data during operation of the analysis tool, thereby keeping input data and analysis results in step.

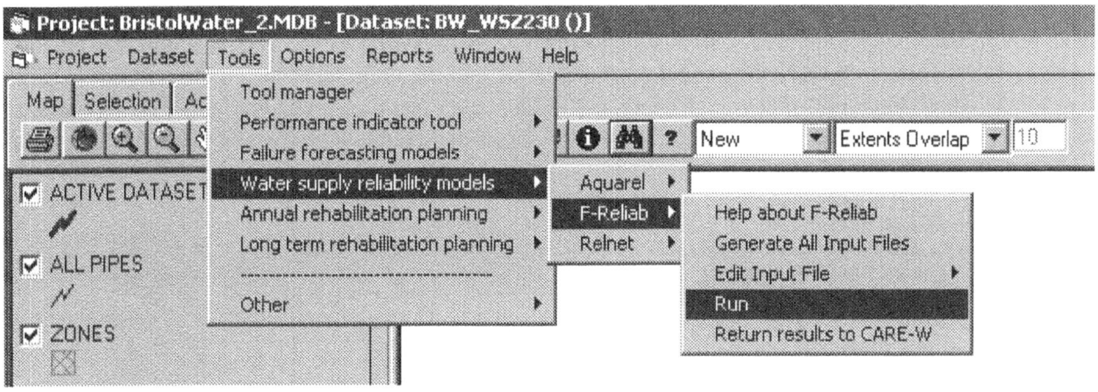

Figure 5.7. Options available to the user when using a rehabilitation planning tool in CARE-W

The GIS viewer is an effective way of displaying results from the analysis carried out by CARE-W tools. Results datasets at individual pipe and dataset (network, sector or cluster) levels can be displayed in separate windows for comparison. A Reports menu allows the user to access all the functions for reporting analysis results. Reports are of two kinds: GIS viewer-based (geospatial) and tabular (Figure 5.8).

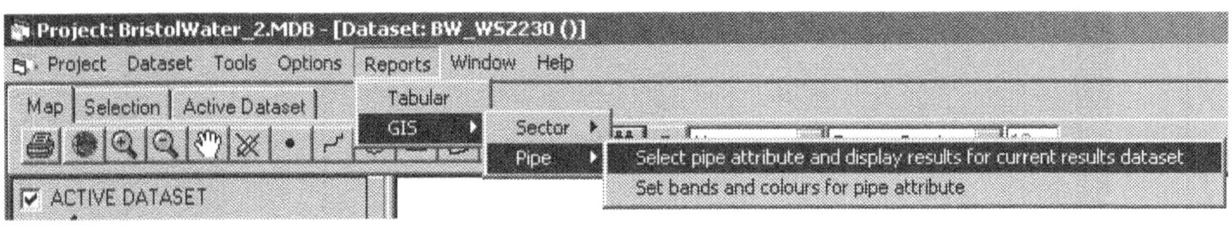

Figure 5.8. Geospatial and tabular reporting options

Two types of results can be displayed in the GIS viewer. The *Pipe* item displays the analysis results for each pipe for the analysis selected by the user, be it from failure forecasting, hydraulic modelling or the ARP tool. The user may choose to view any results relevant to that dataset. The *Sector* option displays results that are not specific to individual pipes such as performance indicators and strategic planning analysis results. To see these results, the user must create dataset boundaries, either manually using the functions on the GIS toolbar, or by selecting the *Automatically generate polygon* command from the Dataset menu.

In addition to geospatial reports, a variety of built-in standard tabular reports are available (Figure 5.9). These combine various sets of similar information available from a number of tools, allowing comparison of analysis results.

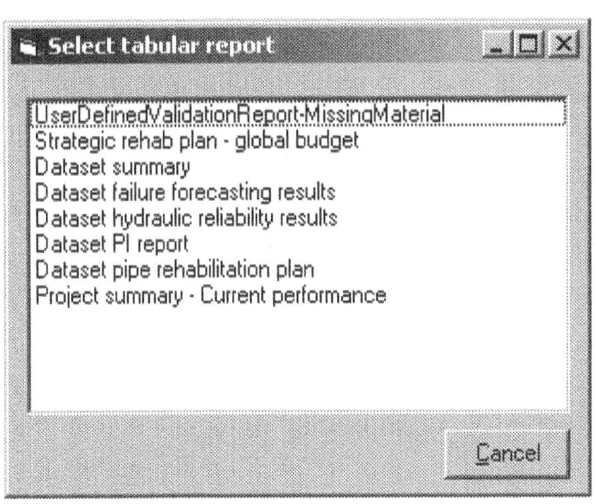

Figure 5.9. Built-in tabular reports

Tabular reports are a valuable addition to any geospatial display (and essential if no geospatial data are available). As they are produced using MS Excel, a permanent record can be kept of the analysis undertaken and the results obtained simply by saving the spreadsheet with an appropriate name. The user is then free to manipulate the data and further spreadsheet analysis may be undertaken if desired.

Figures 5.10 and 5.11 demonstrate the reporting functionality in geospatial and tabular forms. The user may sort and filter tabular report data using MS Excel's own functionality. CARE-W allows the user to tailor the results display by providing an interface to modify the band colour and label, and the range of values in each band. Figure 5.10 shows an example, which highlights pipes at a high risk of failure (in excess of 1 predicted failure per kilometre per year). The parameter (and its unit of measurement) is written out in the results window title bar, together with the dataset whose results are displayed.

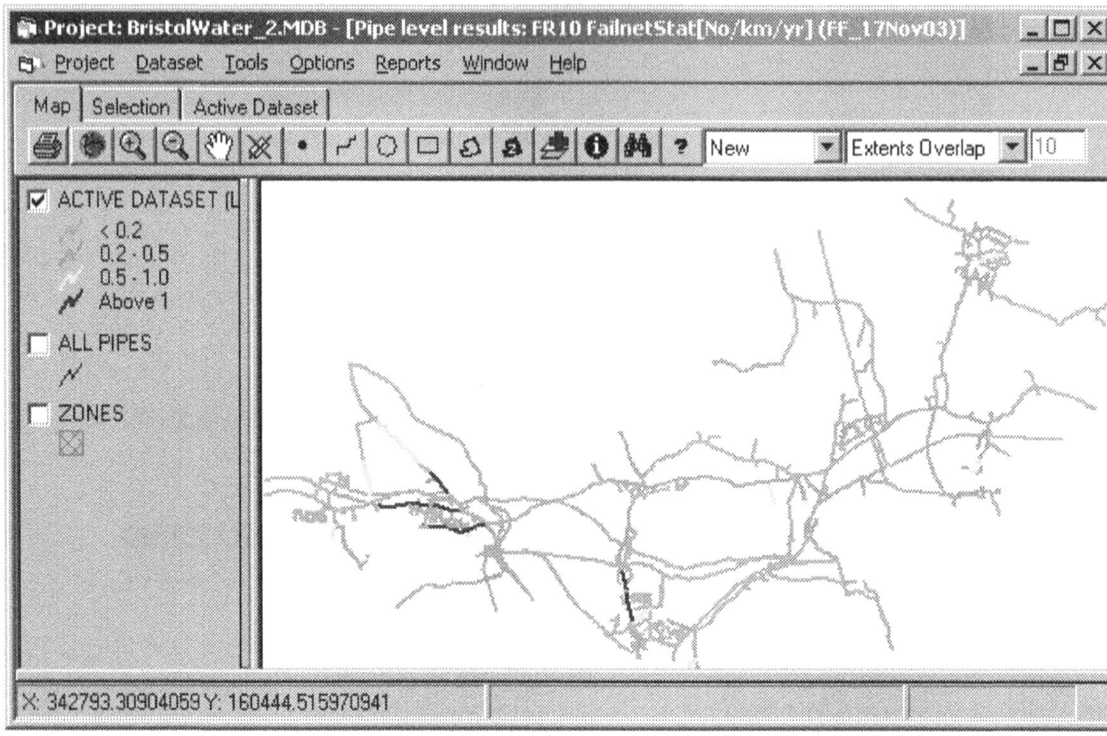

Figure 5.10. Visual targeting of key assets using failure forecasting results

Failure Forecast Results Summary

Project: BristolWater_2.MDB
Active Dataset: BW_WSZ230
Dataset type: Sector
Results dataset: FF_17Nov03

To sort by any field, select all rows except the header rows and use the Sort function in the Tools menu

Failure Rate

User Ref	Main characteristics		Total length (m)	PFR_{10} No/km/yr	Conf. Intervals		FR (Poisson) No/km/yr
	Material	Diameter (mm)			PFR_{10_a} No/km/yr	PFR_{10_b} No/km/yr	
9616114	CI	76	632.34	0.15	0.00	0.63	0.16
9661823	Unknown	76	458.73	0.16	0.00	0.87	0.03
9665348	CI	102	223.09	0.10	0.00	0.89	0.16
9665349	CI	102	32.41	0.13	0.00	3.09	0.16
9665356	AC	100	108.85	0.35	0.00	1.84	9.19
9665357	AC	100	98.57	0.15	0.00	1.01	0.14
9665358	CI	50	136.92	0.29	0.00	1.46	0.16
9665361	AC	100	50.10	0.15	0.00	2.00	0.14
9665362	CI	100	3.41	0.21	0.00	0.00	0.16
9665372	CI	75	615.39	0.20	0.00	0.81	0.16
9665374	CI	75	3.76	0.21	0.00	0.00	0.16
9665375	MDPE	70	2.03	0.20	0.00	0.00	0.02
9665377	AC	100	353.85	0.13	0.00	0.57	0.14

Figure 5.11. Excerpt from failure forecasting tabular report

5.4 CONCLUSION

Practical application foreseen

The Rehabilitation Manager is the means of controlling and managing data flows between tools and is an important tool for selecting data sets and for presenting results. As such, it is an integral part of the tool kit and it is therefore foreseen that the Rehabilitation Manager will have a practical application for most integrated applications of the CARE-W toolkit.

For each potential integrated application of the CARE-W toolkit, the Rehabilitation Manager will support or be used for the following:

- Selection of the pipes that will be analysed by the CARE-W tools;
- Displaying and reporting results of the analysis;
- Allowing user interaction with a pre-determined range of tools under the CARE-W umbrella;
- Interacting with a pre-determined range of tools at the user's request by:
    - creating the necessary input file(s) from the data stored in the CARE-W database to enable the tool to be run, and
    - accepting and storing necessary output data on completion of tool use.
- Providing a central storage area and reference point for essential CARE-W data;
- Accepting **user-validated** input data in a pre-defined format. Data source fields will be mapped to fields within the CARE-W database according to the specification of the declared input data source;
- Providing automatic conversion of data item units to standard CARE-W convention;
- Performing rudimentary data validation including duplication, data type and range checking;

5.5 REFERENCES

Conroy, P: Using CARE-W to manage your water distribution assets, Proceedings of CARE-W Euroconference, Bath November 2003.

Hulance, J: Rehabilitation scheme developer, Proceedings of CARE-W Euroconference, Bath, November 2003.

Kowalski M, Hulance J, Conroy P: The CARE-W rehab manager: EVK1-CT-2000-00053, report D12

Kowalski M, Taylor K, Hulance J, Conroy P, Nascetti D: Elaboration and validation of the CARE-W prototype. The architecture of the CARE-W prototype Software EVK1-CT-2000-00053, report D13

Røstum, J, Kowalski M, Hulance, J: User manual CARE – W Rehab manager. EVK1-CT-2000-00053 report D14, Jan 2004.

Le Gauffre P, Laffrechine K, Schiatti M, Baur R.: Identifying Priority Projects for Annual Rehabilitation Planning. Proceedings of CARE-W Euroconference, Bath, November 2003

Volta M, Poulton M, Herrero M, Røstum J, Sægrov S: Testing and validation of CARE-W Rehab manager. EVK1-CT-2000-00053 report D16, Jan 2004

WRc Publications: WRc Water Mains Rehabilitation Manual: Planning the Rehabilitation of Water Mains, 1989.

6

Testing and Implementation of CARE-W

Jon Røstum, Axel Kønig, Frøydis Sjøvold, Matthew Poulton and Manuel Herrero (NTNU/SINTEF), Marcella Volta (SINTEF/TU Ferrara), Marco Magnioli (TU Bologna), Marcello Schiatti (AGAC), Petra Vrbkova (TU Brno), Joanne Hulance (WRc), Graham Firth and Selina Ward (Bristol Water)

6.1 OVERVIEW OF THE TESTING

6.1.1 Aim

This Working Package has been broken down in two tasks as follows:

Task 6.1. System for testing

The toolkit was tested for 13 real cases reflecting different European conditions, including geographical location, size and water management organisation.

This testing has been carried out in three steps, according to the *CARE-W Description of Work* document:

1. Identification of the current state-of-art
2. Comparative studies of programmes for pipe failure forecasting and water supply service reliability
3. Testing, including evaluation, of the *Rehab Manager*.

Task 6.2. Analysis

A synthesis of the case studies is extracted from the testing to support further business development and implementation of the *CARE-W Rehab Manager*. This synthesis comprises a critical view of the tools and methods and will also include sensitivity and feasibility analyses.

Table 6.1 gives an overview of the test cities, including some properties of water network and management.

The testing phase carried out in this penultimate work package involves all partners and end-users involved with the project The focus is on testing tools within the *CARE-W Rehab Manager* environment rather than as stand alone software.

For most of the test cases there is a end-user report providing feedback and suggestions for further improvement of the software. The individual test reports from the end-users are given as appendices of this document.

 CARE-W edited by Sveinung Saegrov. ISBN: 1843390914. Published by IWA Publishing, London UK

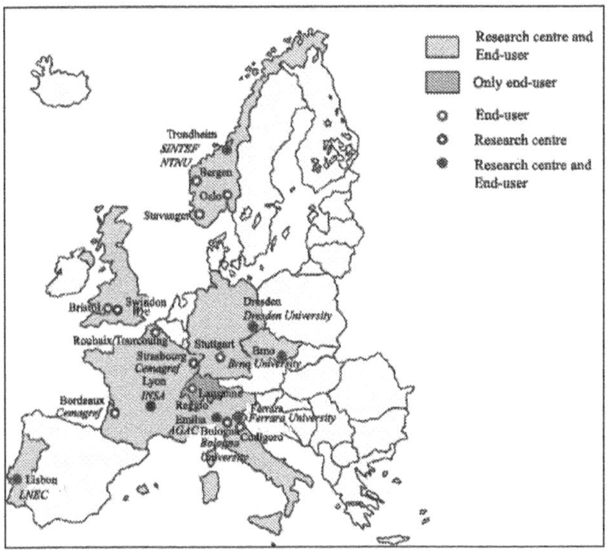

Figure 6.1. Map showing CARE-W partners and end-users (testing sites)

Table 6.1. Properties of the cities

City	Bristol	Brno	Codigoro	Dresden	Ferrara	Lausanne	Lyon	Oeiras/ Amadora	Oslo	Reggio Emilia	Roubaix/ Tourcouing	Stuttgart	Trondheim
Nationality	UK	CZ	I	D	I	CH	F	P	N	I	F	D	N
Location in Europe	W	CE	S	CE	S	C	C	SW	N	S	W	C	N
Population	L	M	S	M	S	S	L	S	M	S	S	M	S
Type of entity	P	P	M	P	M	M	P	M	M	P	P	P	M
Tourist area	N	N	P	Y	N	N	N	Y		N	N	N	N
Data records length (years)		5	6	5	6	50	5-18	30	25	22	90		10
Electronic records	Y	Y	N	Y	Y	Y	Y	Y	Y	Y	Y	Y	Y
GIS	Y	Y	N	(Y)	(Y)	Y	Y	Y	Y	Y	Y	Y	Y
Network length (km)	6400	1100		1600	400	750	3000	750	2000	633	490	1500	700
Inefficiency of use of water resources (%)		20		20		20	10	32	30	22	23	8	20
Water shortage	N	N	N	N	N	N	N	N	N	N	N	N	N
Average pipe age (years)		40		55		28	40	40	30	30	40	35	30
Annual renewal rate (%)		0,5		3		1,3	1	5		1,4	0,6	1	3

Explanation:

1. Location: S,E,N,W,C= south, east, north, west, central
2. Population: L (large, over 1,0 million), M (medium, 0,3 to 1,3 million), S (small under 0,3 million)
3. Type of entity: P=private, M=municipal
4. Tourist area: Y=yes, N=no, P=partly
5. Electronic records: Y=yes, N=no
6. GIS: Y=yes, N=no
7. Water shortage: Y=yes, N=no

The size of networks (or test zones within the network) varied considerably. In some cases it was possible to use the whole network for one or more of the tools, whilst in other regions, small zones within the network, already identified as problematic, were tested. This variety was important in assessing the suitability of each tool given the number of pipes present. For example, failure prediction tools require as many pipes as possible (i.e. a suitably large sample) to ensure statistical significance, whilst large networks will lead to long computational times for hydraulic reliability models.

Table 6.2. Summary of main testing sites including tools tested

Test Network	Size of network / main test zone (km)	CARE-W PI	CARE-W FAIL	CARE-W REL	CARE-W ARP	CARE-W LTP
Bristol, UK	150		✓			
Brno, Czech Republic	1200	✓		✓	✓	✓
Codigoro, Italy	85		✓	✓		✓
Dresden, Germany	11				✓	✓
Ferrara, Italy	33		✓	✓		✓
Lyon, France	1100	✓	✓		✓	
Oslo, Norway	8	✓	✓	✓	✓	✓
Reggio Emilia, Italy	652	✓		✓	✓	✓
Stuttgart, Germany	1420	✓	✓	✓	✓	✓

6.1.2 Testing approach

The testing was carried out by using error/issue log reports. As long as program errors where identified during the testing, partners and end-users were able to report about them. For this purpose, a server was set up with Basic Support for Cooperative Work (*BSCW*) software in order to provide an internal shared workspace for the project for *CARE-W* partners. A standard web form was designed and uploaded to the *BSCW* server intended for that purpose. The users could fill it with all details concerning each error found.

The information required to report any error is:

- Date
- Brief description of issue/error
- Tool where the error first located
- Tool where issue found
- Operating language and system
- Error type
- Contact details

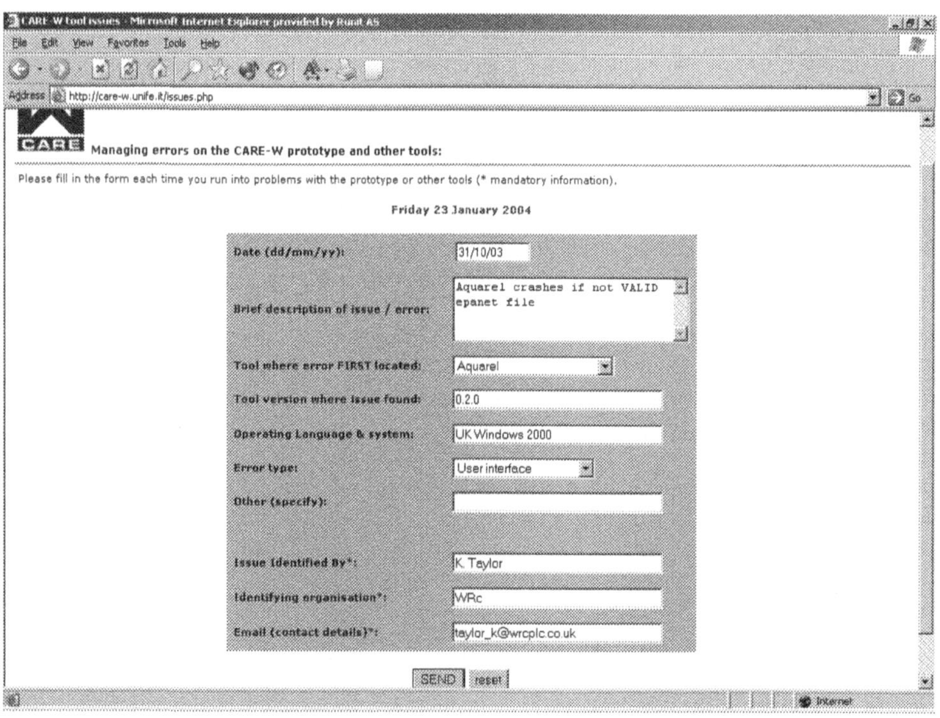

Figure 6.2. The CARE-W issues reporting webfor

All errors reported were stored in a workbook for further investigation (Figure 6.3). Once verified and tested the existence of each error, this was reported to the corresponding tool owner with as much details as possible to make possible to reproduce the error and fix it.

Three possible status of each error were defined following a code of colours:

- Errors still active (red);
- Errors already fixed by the tool owner (orange);
- Errors fixed and validated (green).

The current status of each error was tracked by direct email communication between the chief software developer and the tool owners, updating the issue log as errors were being fixed.

6.1.3 Updating tools

The testing process within the water utilities has been performed by the end-user in close collaboration with partners of the *CARE-W* consortium. End-users could take advantage of the expertise and know-how of technical staff directly involved in the development of the different working packages of *CARE-W*. This made it possible for the users to validate the usefulness of the software within their utilities. Some utilities were concerned about the feasibility of their own data record-keeping systems to work with *CARE-W*. This issue also raised awareness of the lack of data (i.e. short periods of records for failure data) that may be needed to start gathering.

On the other hand, partners got feedback from end-users on the usage of the tools and existing bugs that showed up during the testing. In the mean time, the tools were updated accordingly and uploaded to the *BSCW* server. Descriptions on the bug fixes and any new features added were reported in updated versions of the *Whatsnew.txt* or *Readme.txt* files.

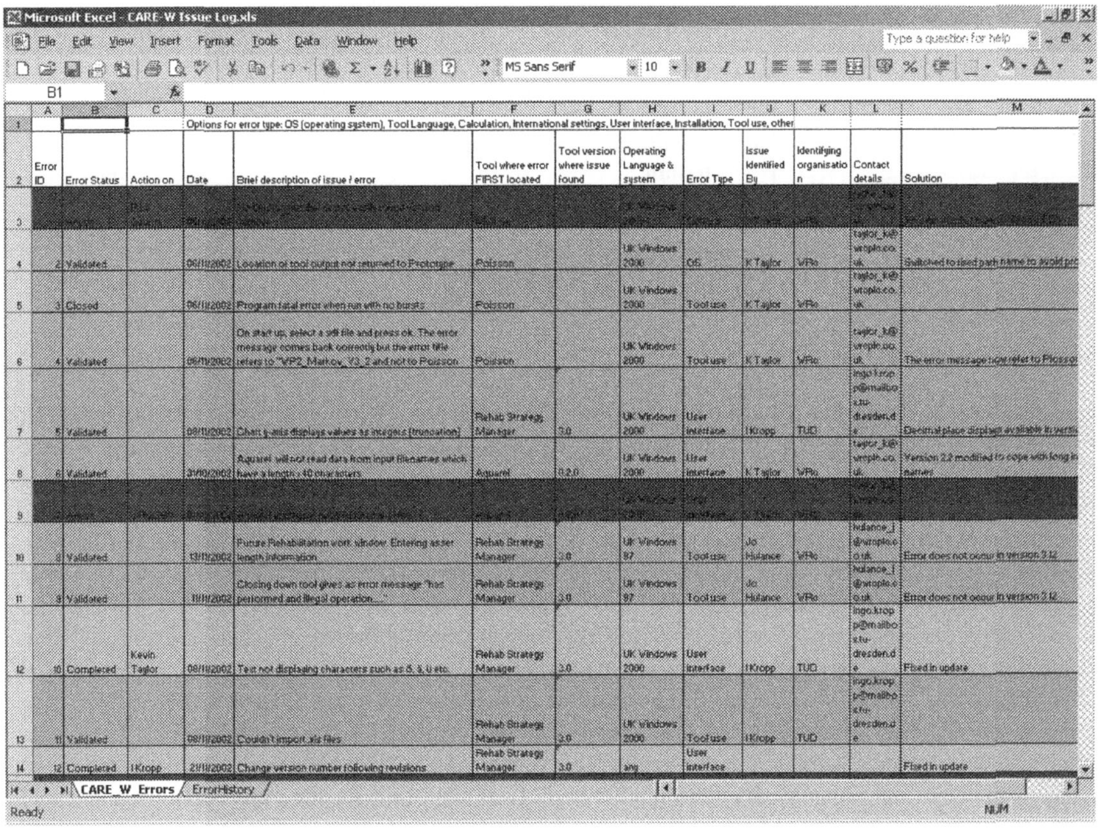

Figure 6.3. CARE-W issue log workbook

6.2 OUTPUT OF THE TESTING

6.2.1 Data availability and preparation

One of the objectives of *CARE-W* is for the software to be applicable to both data-rich and data-lean networks. Data availability influences the way the *CARE-W Rehab Manager* can be used. Some of the *CARE-W* modules are more data-consuming than others. The advanced user with updated GIS, digital maintenance records, calibrated hydraulic network simulation models, etc. can use all the functionality *CARE-W* provides. However, users with fewer data can also benefit, but will more than likely adopt a different path through the software. Exactly which tools to use should be decided case-by-case, based on an evaluation of availability and quality of data.

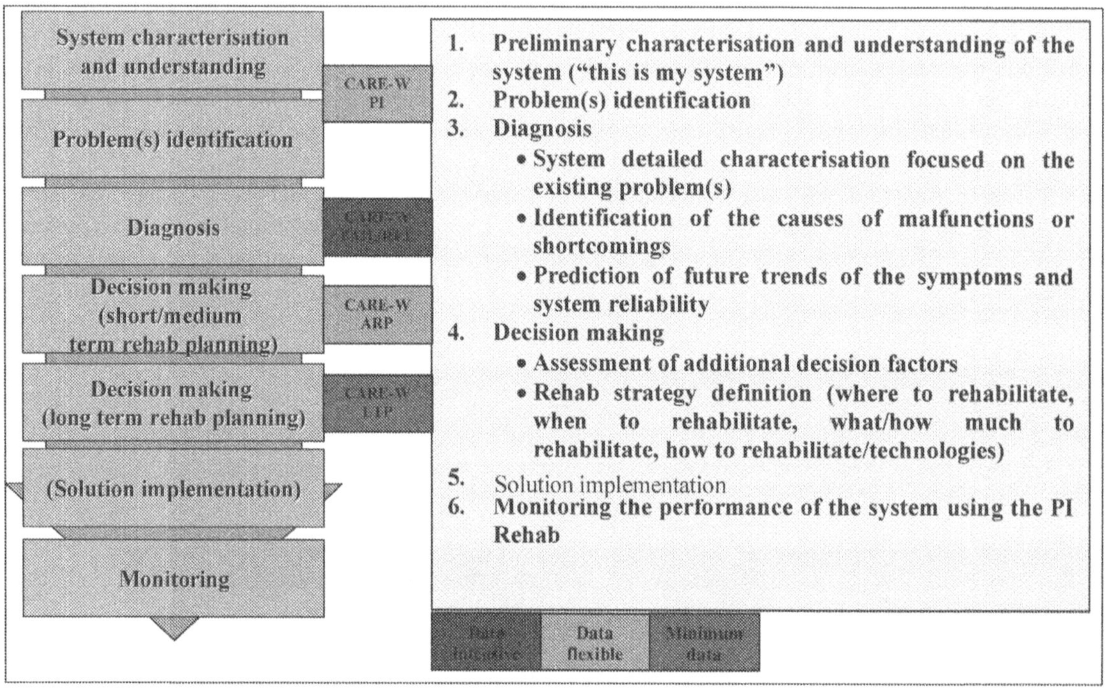

Figure 6.4. Rehabilitation planning procedure

CARE-W can also have the role as a catalyst to start or improve the process of recording data and act as an incentive to store as much data as possible of the highest quality. Tools such as failure prediction models require historical failure records and the variation in record lengths (number of years of data) amongst the test sites, helps determine a minimum recommended value.

In this chapter it is presented the current state of art for the different tools of *CARE-W* resulting from the analysis of the testing. Testing of the toolkit made possible bugs to come out and provided many hints for possible improvements of the *CARE-W Rehab Manager* and the tools for a better integration and a more friendly use of the software. And, vice versa, the applicability of *CARE-W* to real cases in water utilities unmasked obstacles existing on corporate databases to implement the software: lack of data, quality, etc.

Although this testing was mainly focused on the integration of the tools with the *CARE-W Rehab Manager*, individual testing of the tools as stand-alone software could also be performed. In certain situations, the issues shown up on their use through the *CARE-W Rehab Manager*, prevented results from being obtained, so stand-alone use was necessary while bugs were identified, reported and fixed. The state of art for the tools is outlined in Table 6.3.

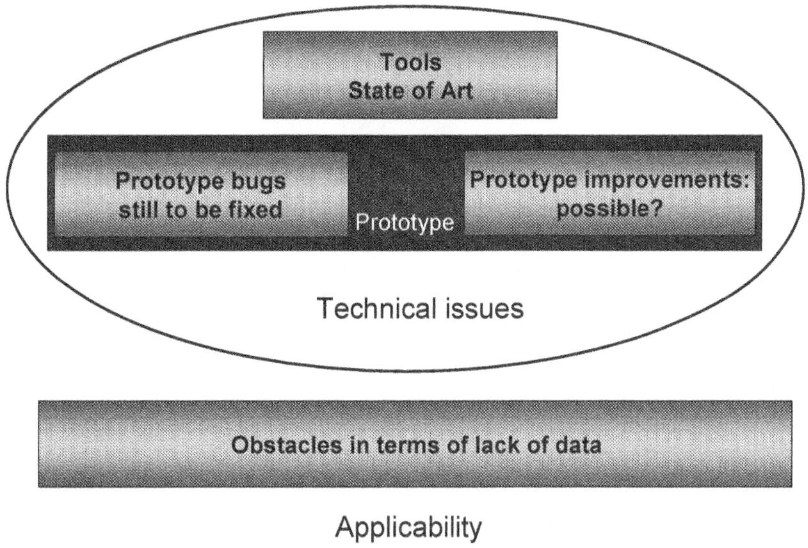

Figure 6.5. Overview of the testing

Table 6.3. State-of-art of the CARE-W tools

Tool	Still bugs	OK as stand-alone	OK in the Rehabilitation Manager
CARE-W_PI Tool	-	Yes	Yes
CARE-W_REL AquaRel	Yes	-	-
CARE-W_REL RelNet	-	Yes	Yes
CARE-W_REL Failnet-Reliab	Yes	-	Yes
CARE-W_FAIL PHM	Yes	-	Yes
CARE-W_FAIL Poisson	-	Yes	Yes
CARE-W_ARP	-	-	Yes
CARE-W_LTP	-	-	Yes

In general, the preparation of raw data to supply the software in order to be able to run is still an important aspect of the *CARE-W* and represents the major workload of *CARE-W* projects (Figure. 6.6). On the other hand, this is maybe the most critical part of the project. In fact, results from *CARE-W* will depend on the quality of the input data: any analysis undertaken using the *CARE-W Rehab Manager* will only be as good as the data used. Therefore, in order to produce reliable results the input data has to be checked.

The *Rehab Manager* in its present version does not supply features for automatic quality control of input data. It is the user's responsibility to ensure that valid data are available for import into the *Rehab Manager*, i.e. the data are arranged in a consistent and definable manner, data types are correct, and data values are meaningful. This should be done before loading the data into the *CARE-W* database. Everyday use and database familiarity is the best way to quality control existing data.

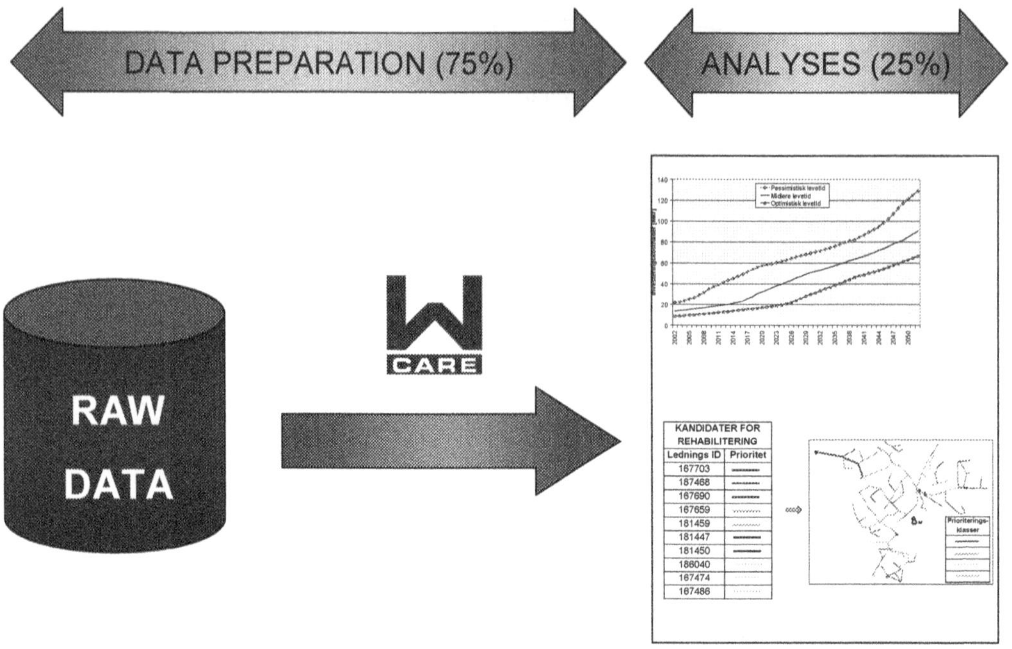

Figure 6.6. Workload in CARE-W

A common data problem is a mismatch between the links in a hydraulic model and the pipes digitized in the GIS. For the reliability tools, values are only calculated for links, which have a corresponding ID in the *CARE-W* database. Thus, it is recommended that the hydraulic model is identical to the GIS network, which is not always the case (e.g. simplified hydraulic models). Additionally, before the analysis starts, failures/breaks must be linked to the pipes where they actually occur. If they are not, the failure tools are not able to run.

The most common known data issues are as follows:

- Breaks not linked to pipes (in some databases the breaks are linked to properties).
- Illogical combinations of pipe material, diameter and installation year (e.g. 150 mm PVC laid in 1930). Normally the diameter is more certain than material and installation year.
- Illogical failure types (e.g. corrosion/rust on plastic pipes).
- Missing pipe data (e.g. material, year, length or diameter is missing). The local database should if possible be updated and corrected. As an alternative, default values might be used, but it is probably best not to include these pipes in the *CARE-W* database and in subsequent analysis, until anomalies have been resolved.
- Failure date pre-dates year of installation.
- Multiple breaks at the same pipe in the same day.
- Rehabilitation date pre-dates year of installation.

In order to cleanse the database and update several values at the same time the *Expression Builder* function in the *CARE-W Rehab Manager* can be used. Here, it is possible to update values for a set of pipes, e.g. all pipes with diameter, 133.3 mm could be changed to 130 mm.

To implement automatic validation checks in the *CARE-W Rehab Manager* is an extensive work and is not included in the present version. However, in the future when *CARE-W* is commercial software such features should be implemented.

Other issue regarding data preparation was the suitability of using rehabilitation entities or clusters rather than single pipes, i.e. groups of pipes with similar attributes, allowing a better planning of rehabilitation activities, based on a zone-wise criterion (topography, system of streets, etc.) that would have not been suitable to undertake if the plan had been developed at single pipe-level. CARE-W does not include a clustering feature therefore extra preparation of data was necessary before running the software.

6.2.2 Main findings

In general, no problems were encountered during the installation of the individual tools. The installation procedure of the *CARE-W Rehab Manager* was not found however particularly straightforward, in part due to the need to set up folders and ensure that files are correctly located within them. However, installing subsequent upgrades is much easier and there is no need to remove and reinstall the software longer.

There are a wide variety of database structures being currently used by utilities. Moreover, structural and maintenance data use to be stored in separate databases. In the *CARE-W Rehab Manager*, GIS asset and failure data are imported as shape files (**.shp*). Some end-users use GIS software that does not produce shape files (e.g. *SmallWorld* by Bristol Water plc), being therefore necessary to extract data with other external tools in order to produce shape files importable by the *CARE-W Rehab Manager*. Nonetheless, several features like selection of elements in the GIS interface, expression builder and help file have proved usefulness.

Since not every tool could be tested completely by all end-users, a procedure was performed in order to evaluate the applicability of each tool to their particular situation. Firstly each tool was rated regarding data availability and benefit for the utility, and thereafter any tool was applied where there was room for profit.

CARE-W_PI

Regarding the individual tools, the *CARE-W_PI Tool* allowed to retrieve a complete outline of the network performance and conditions, disclosing needs for rehabilitation activities.

PI availability on water utilities highly depends on recording habits. End-user have their own data recording system which may allow for the calculation of some indicators in the *CARE-W_PI Tool*, It proved to be useful for finding network zones with potential for rehabilitation and to monitor its effects. The *CARE-W_PI Tool*, together with the *CARE-W_FAIL* tools and the built-in GIS system on the *CARE-W Rehab Manager* have provided end-user with suggestion on the methodology to collect data.

CARE-W_FAIL

PHM and *Poisson* could be tested in several networks, although results could not be properly validated, since data were too few to produce reliable results. Consistency on input data for the program (maintenance and structural data files) is crucial when running this program. Missing input data is an important issue on this program, being necessary to establish a procedure to "fill the gaps" in order to make the program work.

CARE-W_REL

Setup and run of *CARE-W_REL RelNet* encountered no problems and was found to be easy and straightforward. The user-friendly interface made possible an intuitive usage of the tool. However, remarks on the long processing time needed to get results has been pointed out in several cases.

CARE-W_ARP

Data on failures and reliability provide sound criteria for the *CARE-W_ARP* module. These data could be used for the annual planning whenever available, which was not always the case (i.e. network hydraulic model in another file format than *EPAnet*). It was found that in most cases, slight alterations on the criteria used weights chosen by the user did not have a significant effect on the overall results. However, although *CARE-W_ARP* could ease the planning of rehab activities, it is far form replacing staff expertise and knowledge on their networks. It should be used therefore as an aid tool for decision support, rather than decision-making.

CARE-W_LTP

Assessments on long-term rehabilitation needs allowed to establish rehab priorities on specific areas of a network, resulting in up to 50% reduction of failure rates in some cases. However, special care had to be taken when planning annual rehabilitation policies: if socio-economical issues are taken into account as criteria for annual rehabilitation plans, results might significantly differ from those obtained if only hard facts such bursts and leakage are considered.

Uncertainty handling in decision making for long-term plans could be approached by *CARE-W_LTP Scenario Writer* defining upper an lower boundaries for the condition of the water utility as consistent scenarios. This tool proved also to be very usable and intuitive, despite the tough preparatory task of gathering all data needed for the definition of key factors.

Some difficulties arose with the use of the *CARE-W_LTP Rehab Strategy Manager* regarding the assessment on efficiencies on rehab methods. Some of the data needed to define rehabilitation strategies were not available and had to be estimated, thus reducing the confidence of results.

In general, *CARE-W* is seen by some end-users as a complement for existing tools being used already by the water utilities on their asset maintenance plans. For this reason it is critical that data can be imported from or exported to these tools in an easy way. For other end-users, it assist on the identification of goals, providing a standard methodology for rehabilitation planning.

CARE-W usage by the end-user might be a tough task to undertake if there is no additional consultation support from the partners. This is an important finding highlighted several times on this testing: proper training is therefore essential.

6.2.3 Updated help files

Each of the individual tools, and also the *CARE-W Rehab Manager* has its own help system. The testing proved the need of end-users for a suitable on-line help system that makes possible an easy usage of the software and a better understanding of the scientific fundamentals behind each individual tool. Windows Help files created during the first stages of development of the tools have been kept updated according to the suggestions made by end-users and when new features have been added to the software.

The quality on the Help system varies between the different tools. Some are more clear, or user-friendly, than others. Browsing the different help topics is addressed in different ways depending on the tool: through navigation buttons or by means of index of contents, or searching facilities. The need of a common help system format for all the tools has been pointed out.

6.2.4 Bugs

The installation procedure of the *CARE-W* toolkit as a whole has been found to be rather difficult to carry out. Several adjustments on the operating system may be required to run properly, making it necessary to install the software on a separate computer.

Poisson could not successfully conduct an analysis. Further investigation determined that the intermediate text files were not being created when the *mdf* and *sdf* were read by *Poisson* (e.g. mat.txt; ld.txt; diam.txt). It is believed that this issue is concerned with the method used by *Windows XP* to read text files. *Windows XP* is a much higher specification than that agreed by the *CARE-W* consortium. Therefore *Poisson* would not have been designed or tested to be run on this platform.

Special care should be taken on the path for the files used by *CARE-W_ARP* and *CARE-W_FAIL Failnet-Reliab.* These tools do not run properly if input files are not located on the right folder. The help system of each of these tools must provide the user with information on this issue.

Although not a bug, it has been found that *CARE-W_REL RelNet* creates many temporary files during data processing before obtaining results on pipe reliabilities. Those temporary files remain on the hard drive even after closing the tool, requiring therefore manual deletion by the user. Furthermore, the three-step calculation process in *CARE-W_REL RelNet* is not automated, which might be interesting when analysing the reliability of large networks, which require long time data processing. These two latter are main issues specially in resource-limited computers. These issues have been recently addressed in the release 2.04, together with some improvements on stability and system resources consumption.

Some problems arose when producing the input file for *CARE-W_ARP* with the *CARE-W Rehab Manager*. Some data were missing and had to be added manually to the input file.

In some GIS systems, each pipe consists on multiple elements, depending on the digitisation method, having all elements the same user reference. However, the *Rehab Manager* would treat each GIS element as an individual main and therefore would not be able to import more than one element with the same user reference. This has been fixed after version 1.5.2.

Regarding the use of the *CARE-W_LTP* tools, data transfer from the *CARE-W_LTP Scenario Writer* to the *CARE-W_LTP Strategy Manager* and *Rehab Strategy Evaluator* has not been fully understood by some end-users, so testing was not possible.

A few bugs found when defining import specifications (*definitions for data transfer*) have been fixed in version 1.5.2 of the *CARE-W Rehab Manager*, improving and easing the use of the tool.

The *CARE-W_REL RelNet* input file (i.e. an **.inp* file) produced by the *CARE-W Rehab Manager* is not always fully compatible with *CARE-W_REL RelNet*. In certain cases, the original input file from the *EPAnet* model has been necessary to be used in order to make the calculations. Although some tolerance to input files supplied by the *Rehab Manager* has been implemented in *CARE-W_REL RelNet*, there might still be some room for improvement of the compatibility within *CARE-W Rehab Manager* and *CARE-W_REL RelNet*.

6.2.5 Proposals for enhancement

The maintenance and upgrading (i.e. compatibility with *Windows XP*) of the *CARE-W* software and toolkit is an issue that needs to be addressed in the commercial phase of *CARE-W*, particularly as Microsoft is continuously upgrading their systems.

Some tools use external software (e.g. *CARE-W_REL RelNet* and *CARE-W_REL AquaRel* use *EPAnet2* dynamic link library, and *CARE-W_LTP* tools use *FireBird*). As long as *CARE-W* results rely to certain extent on that software, compatibility should be maintained for future versions of this.

Extra functionalities on the *CARE-W_LTP Rehabilitation Strategy Manager* have been suggested by end-users, namely comparison between strategies within the same window or saving of settings when closing the program.

Further development of the *CARE-W_LTP Scenario Writer* and *CARE-W_LTP Strategy Evaluator* could be addressed towards its applicability in utilities providing other services, or even multi-services utilities, in which long-term planning becomes more relevant since some key factors may influence more than one service at once.

Some tools, e.g. *CARE-W_REL RelNet*, still require much computer resources to run. Code optimization towards resources consumption should be addressed in future releases. Capability to work with entities larger than pipe sections is also to be implemented in future versions of *CARE-W_REL RelNet*.

It would be beneficial if the *Tool Manager* of the *CARE-W Rehab Manager* had a browse facility to ease the selection of file locations. Furthermore, it might be interesting to have this selection automatically done by default after the installation *CARE-W Rehab Manager* has finished, so the user will not need to care about this issue.

It might be needed to agree on the requirement of the *CARE-W Rehab Manager* to be able to import at least, certain standard interchange GIS file format usable by most of the GIS software packages existing on the market. Furthermore, it may be able to import any other proprietary formats.

Other suggestions to improve the use of the *CARE-W Rehab Manager* have been made by the end-users, such as the use of the *Expression Builder* feature, sorting of table records and resizing of windows within the interface. On the other hand, the example project file supplied with the software (*Crissier.mdb*) lacks some structural data, which should be amended, so end-users will be able to learn the use of *CARE-W* without trouble.

A better integration between the different tools and the *CARE-W Rehab Manager* involves that when any tool is launched from the latter, it could "remember" the files (name and path) to be processed. The same may apply when returning processed results to the *CARE-W Rehab Manager*.

Other minor improvements, although not critical, might be considered to be included within the software. Better integration with the Windows environment can be achieved by using controls from the Microsoft *Common Dialogs* dynamic link library in many of the forms within the tools. This is applicable to most of the tools so there is still some room for the improvement of the interface, reducing the opportunity for user error, providing all the tools of *CARE-W* with a similar appearance.

Some users appreciated also the possibility to have *CARE-W* translated to the local language. At the current stage, this is only possible in *CARE-W_LTP Rehabilitation Strategy Manager*.

6.3 CONCLUDING REMARKS

Overall, the tools included in *CARE-W* can work successfully with the *CARE-W Rehab Manager*. All bugs identified during the testing have been fixed and validated. Compatibility between input and output data from *CARE-W Rehab Manager* and the individual tools has been found to be an important issue, and proved to be acceptable at the current stage. Compatibility with existing information systems used by utilities is still an issue to be addressed.

The testing carried out by end-users made it possible to evaluate data availability on corporate databases as well as their structure, thus enabling the way for future implementation of rehab programmes based on *CARE-W*. New ideas for data recording have been identified by the end-users. On the other hand, extra functionality needed in the software was identified.

Feedback provided by the testing and validation of *CARE-W* extends and enhances the current possibilities to approach rehabilitation of water networks. End-user participating in the testing of *CARE-W* generally validated the software as a useful tool that fits with those already being used in water utilities for asset maintenance management. It is expected that the *CARE-W* toolkit as a whole will allow for the definition of a recommended best practices for future rehabilitation plans, based on data availability and management goals.

6.4 REFERENCES

ALEGRE, H.; TUHOVČÁK, L.; VRBKOVÁ, P. 2003. *Performance management and historical analysis: the use of the CARE-W PI Tool by the Brno Waterworks Municipality*. CARE-W International Conference Water Infrastructure Management: Planning Water Mains Rehabilitation. Bath, UK.

ALGAARD, E.; CAMPBELL, P. 2003. *Critical needs for rehabilitation planning*. CARE-W International Conference Water Infrastructure Management: Planning Water Mains Rehabilitation. Bath, UK.

BAUR, R; KROPP, I.; TERNO, O. 2003. *Long term strategic planning: determining rehabilitation investment needs*. CARE-W International Conference Water Infrastructure Management: Planning Water Mains Rehabilitation. Bath, UK.

CONROY, P. 2003. *Using CARE-W to manage water distribution pipes*. CARE-W International Conference Water Infrastructure Management: Planning Water Mains Rehabilitation. Bath, UK.

EISENBEIS, P.; POULTON, M.; LAFRÉCHINE, K. 2003. *Technical indicators for rehabilitation: failure forecast and hydraulic rehabilitation tools.* CARE-W International Conference Water Infrastructure Management: Planning Water Mains Rehabilitation. Bath, UK.

FIRTH, G.S.; WARD, S.V. 2003. *The benefits of integrated planning tools – the end user perspective.* CARE-W International Conference Water Infrastructure Management: Planning Water Mains Rehabilitation. Bath, UK.

HULANCE, J. 2003. *The CARE-W Rehabilitation Scheme Developer.* CARE-W International Conference Water Infrastructure Management: Planning Water Mains Rehabilitation. Bath, UK.

KOWALSKI, M. 2003. *The CARE-W Rehabilitation Manager software.* CARE-W International Conference Water Infrastructure Management: Planning Water Mains Rehabilitation. Bath, UK.

LE GAUFFRE, P.; LAFRÉCHINE, K.; SCHIATTI, M.; BAUR, R. 2003. *Identifying priority projects for annual rehabilitation planning.* CARE-W International Conference Water Infrastructure Management: Planning Water Mains Rehabilitation. Bath, UK.

POULTON, M.; RØSTUM, J.; VOLTA, M. 2003. *Testing the CARE-W toolkit.* CARE-W International Conference Water Infrastructure Management: Planning Water Mains Rehabilitation. Bath, UK.

SÆGROV, S. 2003. *What is CARE-W?* CARE-W International Conference Water Infrastructure Management: Planning Water Mains Rehabilitation. Bath, UK.

7

Integrated use of CARE-W

Jon Røstum and Sveinung Sægrov, SINTEF

7.1 THE CARE-W PROCEDURE

CARE-W is a computer-based system for water network rehabilitation planning. It consists of software dealing with fundamental instruments for estimating the current and future condition of water networks, i.e. Performance Indicators, prediction of network failures and calculation of water supply reliability. Furthermore it includes routines for estimating long-term investment needs as well as selection and ranking of rehabilitation projects. These tools are integrated and are operated jointly in the CARE-W Rehabilitation Manager, which also contains facilities for use including database, GIS and in- output routines. The results are presented by reports, in tables and by graphics/GIS.

In addition to the tools shown in the CARE-W Toolkit, CARE-W also provides a cost-estimating tool for estimation of costs for rehabilitation of the pipes resulting from the CARE-W ARP module.

7.2 A REHABILITATION PLANNING SEQUENCE

Efficient planning requires the rehabilitation engineer or planner to be in possession of a substantial amount of background knowledge and experience on the types of problems faced, current performance and possible effective solutions. The engineer must be aware of the objectives of rehabilitation for each problem he faces, and apply sound judgement using all the tools at his disposal in an appropriate manner. This places a huge burden on the engineer when many rehabilitation methods are feasible and there are many solutions to improve service delivery to customers.

There are several ways of working with CARE-W depending on the purpose of the analysis and the availability of data. A complete CARE-W rehabilitation sequence can be summarised in Figure 7.2.

CARE-W covers the items typed in **bold** (which mean all the issues, except for rehabilitation technology (partly covered) and the practical implementation of the technology in the ground).

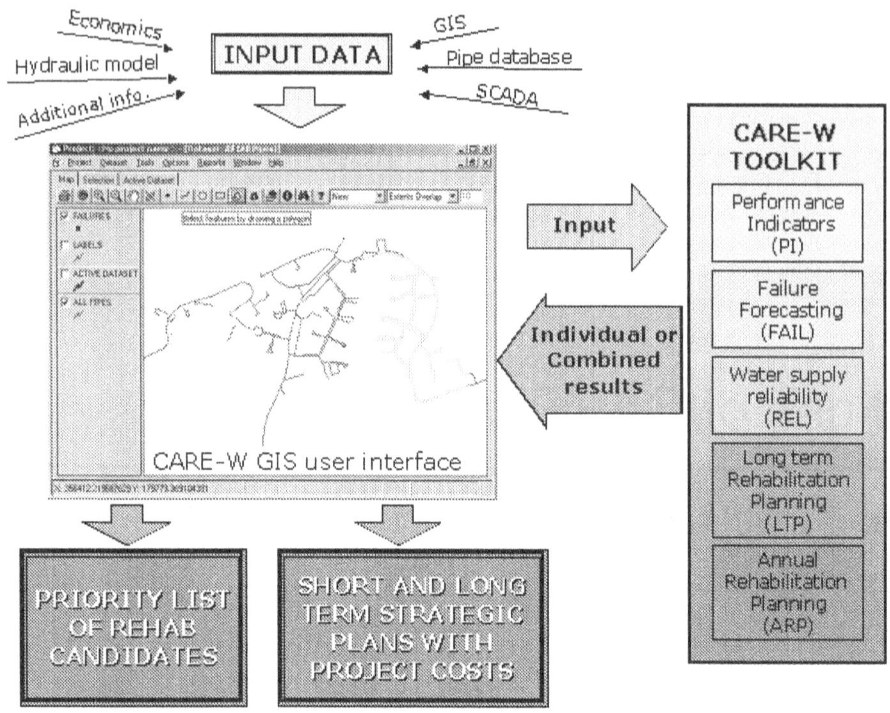

Figure 7.1. Overview of CARE-W

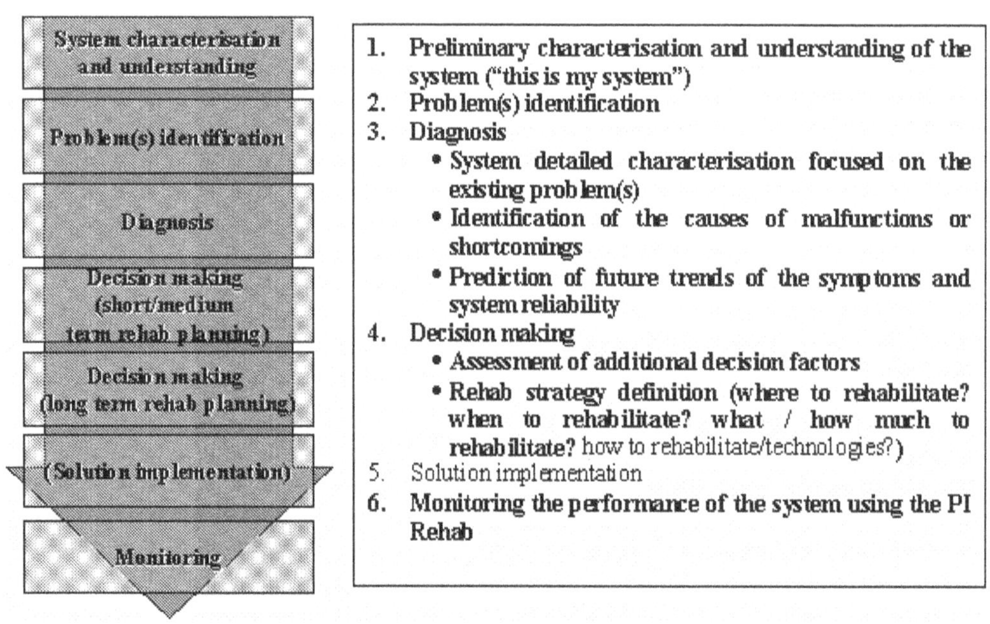

Figure 7.2. CARE-W procedure

How is the system working? What are the problems? Where are the problems?

The CARE-W PI module can be used for preliminary characterisation, understanding of the system and for problem identification. The performance of the system at network, sector and cluster level are identified. The rehabilitation planner can then decide on which zones to investigate in more detail. Monitoring of the system performance continues after completion of the rehabilitation program to confirm that the expected benefits have been achieved and to identify any problem that may develop over time. PI's presented as time series are useful for this purpose.

How much money is needed for rehabilitation in the long- term, 10-20 years? Which long-term rehabilitation strategy is the best?

The CARE-W LTP provides tools for analysing future scenarios, forecasting long-term rehabilitation needs and choosing the best rehabilitation strategy. After applying this module, the rehabilitation planner will know which strategy to choose and the investment needed.

Which pipes should be rehabilitated?

The decision problem for the water utility is how to spend a given annual budget for rehabilitation (resulting from the long-term budget) on the most efficient projects. For selecting the limited number of pipes for the annual rehabilitation program different criteria must be taken into considerations. Generating annual rehabilitation programmes is thus a *multi criteria task*. The CARE-W ARP module uses information from the other CARE-W modules CARE-W PI, CARE-W FAIL and CARE-W REL. Links this with information about other aspects (i.e. criteria). After processing the data the final result is a prioritised list of potential candidates for rehabilitation.

7.3 HOW CARE-W CAN HELP?

The CARE-W Rehabilitation Manager software is a versatile and powerful application, incorporating a variety of rehabilitation planning tools. It has the flexibility to allow the rehabilitation of networks to be planned at both strategic and tactical levels, or in other words, at various levels of complexity: network, zone or pipe. This flexibility is possible because of the nature of the tools that form the Rehabilitation Manager software. It should be noted that the CARE-W Rehabilitation Manager software is a tool to *help* an engineer devise annual and strategic rehabilitation plans; it does **not**, and is **not intended to** produce the plan itself.

7.4 DATA ISSUES

7.4.1 Basic requirements

The data availability influences the way CARE-W Rehabilitation Manager can be used. Some of the CARE-W modules are more data consuming than others. The advanced user with updated GIS, digital maintenance records, calibrated hydraulic network simulation model etc. can use all the functionality CARE-W provides. However, users with fewer data can also benefit from CARE-W. Exactly which tools to use should be decided case by case, based on an evaluation of availability and quality of the data. However, CARE-W can also be use as a catalyst to start or improve the process of recording data and act as an incentive to store as much data as possible of the highest quality.

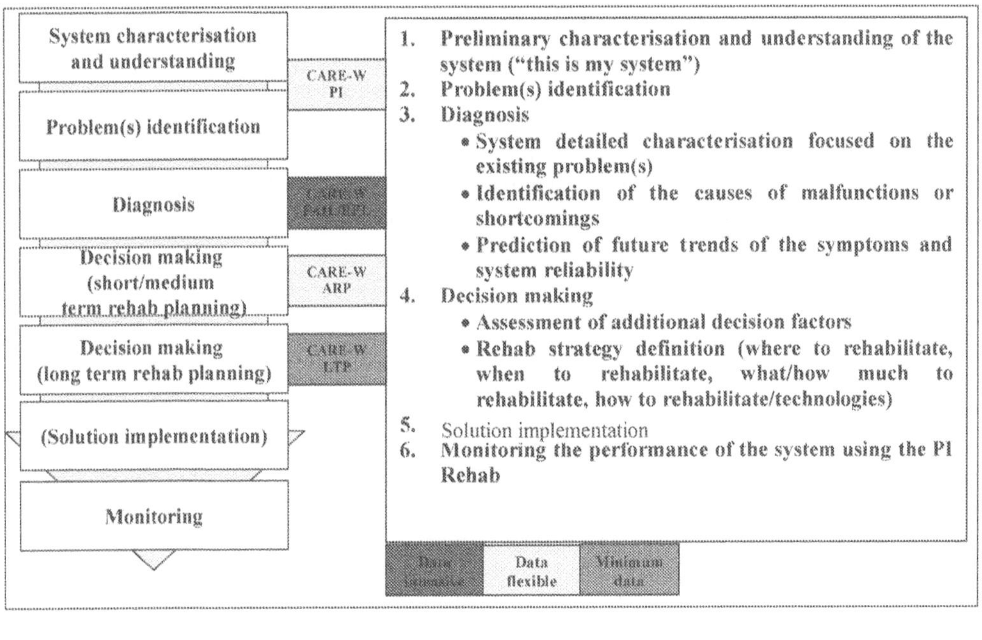

Figure 7.3. CARE-W rehabilitation planning procedure and the data requirements for the different CARE-W tools

CARE-W FAIL requires detailed information about pipe inventory and failure data. CARE-W REL might for large networks (e.g. many pipes) require long simulation times (i.e. hours). CARE-W PI and CARE-W ARP are more flexible with respect to data requirements. Results from CARE-W FAIL and CARE-W REL are normally used as input for CARE-W ARP.

7.4.2 Software basics

The CARE-W Rehabilitation Manager consists of a central MS Access 2000 database and a Visual Basic 6.0 application for the user interface.

A PC with the following minimum specification will be required to run CARE-W and the failure forecasting tools effectively:

- 1 GHz Processor
- 256 Mb RAM
- Windows 2000
- Office 2000 Professional

This is the minimum specification agreed by the CARE-W consortium. It might also run for smaller computers, but then more time is required for loading data, and running analysis etc.

7.5 WORKING WITH PROJECTS AND DATASETS

7.5.1 Projects

A CARE-W project is a collection of data items, analyses and results pertaining to an area or areas of interest to the CARE-W user. The area of interest may be geographic (such as a city network) or thematic (e.g. water quality issues). The notion of a project provides a means to work with, and maintain, multiple CARE-W model simulations.

7.5.2 Datasets

A CARE-W project in use will contain one or more datasets. A dataset is a user-specified collection of pipes. All analysis within CARE-W will be carried out on an input dataset and, for each simulation, a results dataset will be created. Clearly, the link between pipe attribute information and the results from analysis carried out on these pipes must be maintained. An electronic analysis history will be kept, and results datasets named such that they reference the analysis tool and input dataset used. All analysis within the Rehabilitation Manager is conducted on the active dataset, which, as the name suggests, is the current, selected dataset on which the user is working.

Input datasets may be of three types: network, sector, or cluster. Datasets may only belong to one project and are created by selecting pipes. Pipes may be selected for inclusion in a dataset in a variety of ways depending on whether geospatial information is available or not. A *sector* is understood as a collection of pipes which have been grouped under a reference name, typically an operational or management unit, e.g. water supply zone, hydraulic zone, district meter area. A *cluster* is understood, in practice, to mean a geographic clustering or grouping of pipes with one or more common attributes (such as age, material, geographical location, etc.) which may or may not lie within one sector. A *network* is understood to be the largest grouping of pipes, typically a collection of sectors of a company or municipality.

7.6 WORKING WITH CARE-W TOOLS: INTERACTION

A main feature of the Rehabilitation Manager software is the facility to interact with a pre-determined range of tools. CARE-W is not intended to bind together the external tools in a fixed and constraining way, but rather to allow the user to use them individually or in a sequence appropriate to the data available for analysis.

Once the user has selected a tool to use for his analysis, data files need to be created which will allow the necessary data to be used by the tool. The CARE-W Rehabilitation Manager has a facility to import pre-existing input files for each of the CARE-W tools. CARE-W Rehabilitation Manager calculates many of the values required by the PI Tool directly and automatically based on information held in the central database.

Analysis tool input files are created from data previously imported to the relevant CARE-W project database. An additional feature of the Rehabilitation Manager is the facility to allow the user to view and edit the import files created by the Rehabilitation Manager before the tools use them. This allows the user to check the integrity of their contents and to ensure successful file creation. If the input data files are edited interactively by a tool (e.g. CARE-W Fail or CARE-W REL), the edits are returned to the CARE-W project database in addition to the results (output) data.

For detailed help concerning the individual tools see each tools individual Help menu. In the CARE-W help system only a **Brief Help for the tools** is provided.

7.7 GETTING RESULTS IN CARE-W

The results from CARE-W can be displayed either in GIS or tabular reports. CARE-W is flexible enough to be present useful results regardless of whether the water works have a GIS system or not.

7.7.1 Geospatial reporting

Geospatial results are available both at sector level (e.g. leakage zones) and at pipe level. Figure 7.4 shows PIs for different zones. This kind of map is useful when you want to focus on the areas where the problems occur.

Figure 7.4. Geospatial information

The GIS pipe view item will display the analysis results for each pipe (pipe ID) that correspond to results obtained from the asset-based analysis tools (CARE-W FAIL, CARE- W REL and CARE- W ARP), where these data exists.

The sector results (GIS display) will display results that are not specific to individual pipes. These would be results obtained from the CARE-W PI and CARE-W LTP modules. In order to report results at a sector level in GIS the user must manually assign boundaries to a dataset as described under dataset boundaries.

7.7.2 Tabular reports

Under the *CARE-W menu*, by clicking **Reports** the user will have access to different report templates. Results will be produced based on information held in the project database for the current ('active') dataset.

The tabular menu item allows the user to access non-GIS reports, detailing both project summary information and analysis results from individual CARE-W tools. The reports are created by querying data directly from the central database and exporting it to a spreadsheet application (currently MS Excel®). The advantage of using these tabular reports is that a permanent record can be kept of the analysis undertaken and the results obtained by simply saving the spreadsheet with an appropriate name. All the tabular reports can be printed and saved by the user. The data within the reports can also be manipulated by the user and further analysis undertaken, if required. For the advanced user it is also possible to make user-defined reports in MS Access® (mdb query template).

When the **Tabular** reports menu item is selected a dialog box will appear listing the reports that can be generated by the Rehabilitation Manager. This list of reports is shown under the *CARE-W menu.* The following reports are available:

- **Strategic Planning – global budget report.** The aim of this report is to present the strategic budget value (in Euro) and the rehabilitation strategy that has been specified to develop this budget.

- **Project summary – current performance report**. This report summarises the calculated performance indicators (PI) for a given CARE-W project.
- **Dataset summary.** This is a straightforward report, which summarises the current status of the active dataset, i.e. before any rehabilitation plan is implemented
- **Dataset failure forecasting results**. This report summarise asset-based results from the failure forecasting tools, CARE-W FAIL.
- **Dataset hydraulic reliability results.** This report summarise asset-based results from the hydraulic reliability modelling tools, CARE-W REL.
- **Dataset PI report**. This report will summarise the calculated PI values (CARE-W PI) for the active dataset.
- **Dataset pipe rehabilitation plan**. This report presents the analysis results from CARE-W ARP module.

7.8 USING YOUR RESULTS

7.8.1 What do they mean?

The CARE-W Toolkit can be used in many different ways, depending on the users' needs and the data availability. Based on this the user should be able to identify his/her needs with respect to which tools to apply. Figure 7.5 gives additional information about the integrated use of the tools and how to interpret the results.

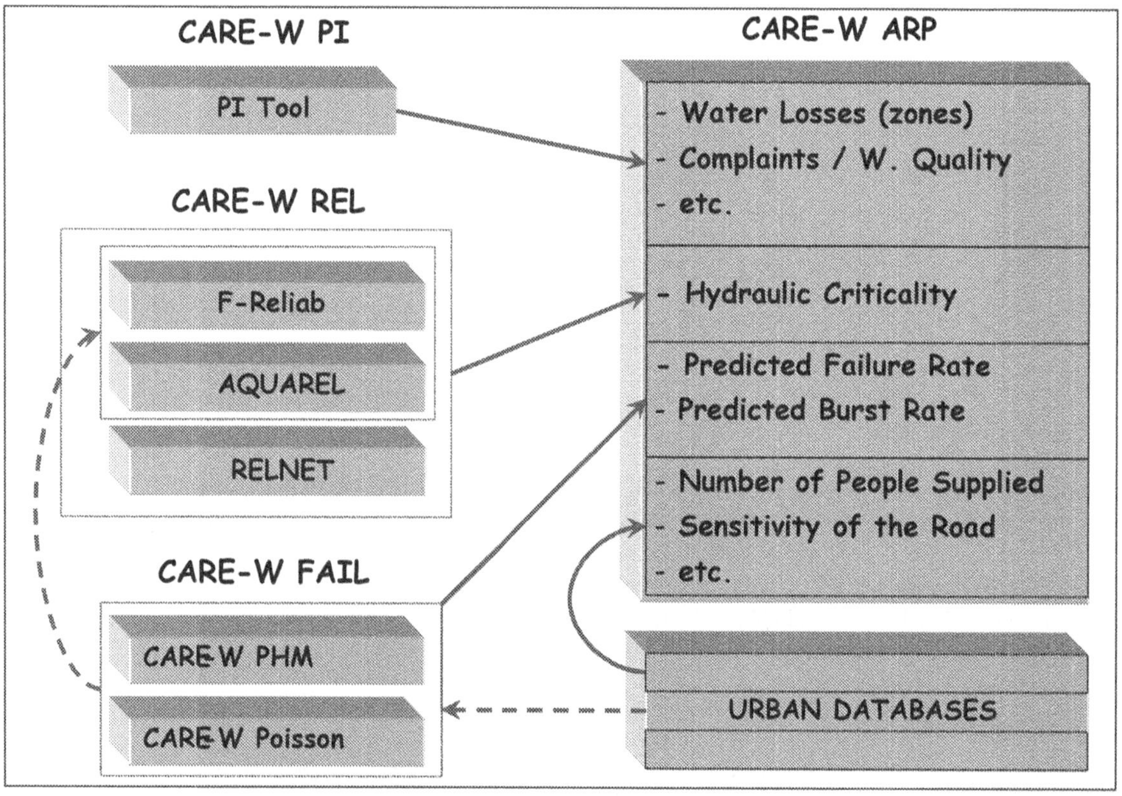

Figure 7.5. Integrated use of the results in CARE-W for generating annual rehabilitation plans

CARE-W PI is a powerful diagnoses and decision support tool. It might also serve the CARE-W ARP module with information for generating the annual rehabilitation plan.

For CARE-W Fail there are two alternative models, CARE-W PHM and CARE-W Poisson for calculating the failure rate for each pipe. The failure rate (FR) for each pipe is a major parameter to calculate CARE-W ARP criteria. The statistical analysis that the CARE-W Fail tools also provides useful information about the behaviour of your system and which factors influencing the failure history of the network.

In CARE-W there are three alternative reliability models (i.e. Aquarel, F-Reliab, and Relnet). The models calculate a *hydraulic criticality index* (HCI), expressing the hydraulic impact/consequences due to a pipe failure in the network. The HCI is one of the possible criteria to be used in CARE-W ARP. Since the different models have different approaches the absolute values cannot be compared. The tools Aquarel and F-Reliab include failure data in the calculations and Relnet does not. Nevertheless, the relative ranking of the critical pipes are normally similar. The user should use the model he is most familiar with.

Special attention should be made for CARE-W ARP and the ranking procedure. Based on different criteria (impacts) the pipes are ranked in six (6) categories (e.g. C33, C32, C31, C22, C21 and C11). The most likely candidates for rehabilitation are in category C33, then follows C32, C31, C22 and so on.

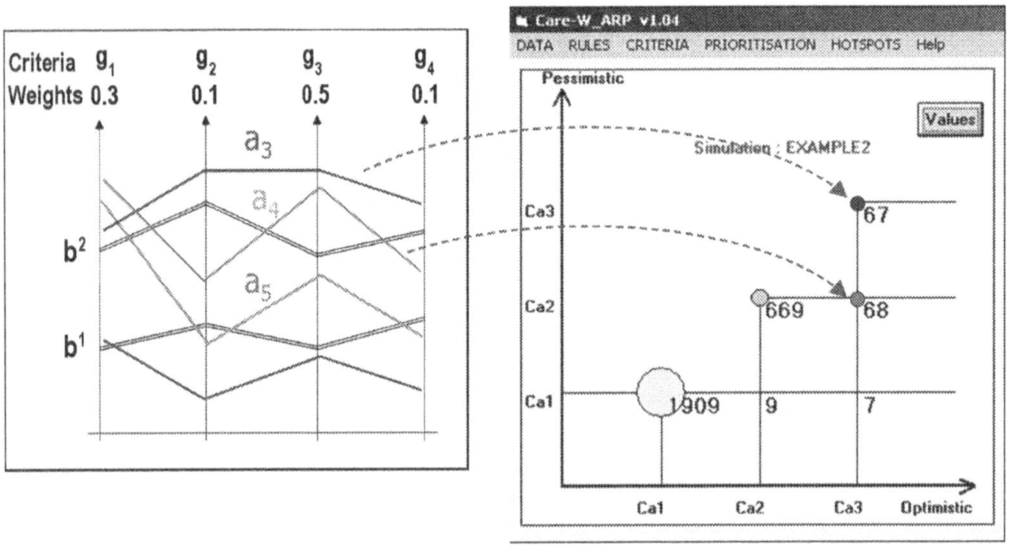

Figure 7.6. Outranking procedure in CARE-W ARP

The values for the weights and thresholds in CARE-W ARP influence the distribution of pipes in the categories and can be used as "adjusting screws".

The *costing tool* picks up results from the CARE-W ARP and for each candidate for rehabilitation (e.g. pipes) the recommended rehab technique, cost, and time for rehabilitation is given.

7.8.2 Supplying inputs for the rehabilitation plan

Results from CARE-W can be used as input for rehabilitation planning in many ways. The individual tools supply useful information for understanding the network. The CARE-W ARP provides support by ranking potential rehabilitation projects to be included in the rehabilitation plan for the water utility.

7.8.3 What now?

It is not the intention of CARE-W that the central CARE-W database should replace the users own database system. An efficient way of working with CARE-W is to import data from the local database, run CARE-W with all or some of the tools and afterwards export relevant results back to the user's local database. Example of useful information to export back might be:

Pipe level:
- Failure rate (CARE-W Fail)
- Break rate (CARE-W Fail)
- Hydraulic criticality index (CARE-W REL)
- Priority level (CARE-W ARP)
- Hotspots (for criteria in ARP).

Cluster/sector/network level:
- PI results (CARE-W PI)
- Long-term rehab needs/strategy (CARE-W LTP)

In this the information provided by CARE-W serves as a basis for infrastructure data management.

7.9 Conclusions

CARE-W has the potential for a comprehensive multi-criteria analysis of the current water network condition and performance, the long-term investment needs and the ranking of rehabilitation projects. The system is to some extent biased towards bursts and quantitative water supply. Water quality water leakage issues are handled via the Performance Indicator System and as criteria in the multi-criteria analysis tool for project prioritisation. These tools can be used in conjunction with monitoring leakage and quality. In the future a more comprehensive integration of advance tools for water quality and leakage is foreseen.

For the time being the CARE-W software is not freeware on the international software market. The commercial use is meant to take place via the developers and trained experts ("CARE-W champions")

7.10 References

Baptista JM, Alegre H: Construction of a control panel of performance indicators for water networks; EVK1-CT-2000-00053, report D1 Dec 2001.

Baptista JM, Alegre H: Validation of the rehabilitation performance indicators systems. EVK1-CT-2000-00053, report D2 March 2002

Eisenbeis P, Legat Y, Laffrechine K. Legauffre P, Kønig A, Røstum J, Tuhovcak L, Valkovic P: Description and validation of technical tools, EVK1-CT-2000-00053, D3

Eisenbeis P: Testing and validation of technical tools, EVK1-CT-2000-00053, D4

Eisenbeis P: Guidelines for the use of technical tools, ECK1-CT-2000-00053, D5

LeGauffre P, Lafreuchine K, Baur R: Decision support for annual rehabilitation programs. Criteria for the prioritization of rehabilitation projects: EVK1-CT-2000-00053, report D6

LeGauffre P, Baur R, Laffrechine K, Miramond M: Decision support for annual rehabilitation programs. Survey of multi-criteria techniques and selection of relevant procedures. EVK1-CT-2000-00053, report D7

LeGauffre P: Description of CARE-W ARP software, EVK1-CT-2000-00053, report D7

Herz R, Lipkow A: Strategic planning and Investment: Development of the "Scenario Writer" software for the exploration of future utility backgrounds EVK1-CT-2000-00053, report D9

Herz R, Kropp I: Strategic planning and Investment. Development of the Rehab Strategy Manager" software: EVK1-CT-2000-00053, report D10

Herz R, Baur R, Kropp I, Lipkow A: Strategic planning and Investment. Development of the "Rehab Strategy Evaluator" software. EVK1-CT-2000-00053, report D11

Kowalski M, Hulance J, Conroy P: The CARE-W rehab manager: EVK1-CT-2000-00053, report D12

Kowalski M, Taylor K, Hulance J, Conroy P, Nascetti D: Elaboration and validation of the CARE-W prototype. The architecture of the CARE-W prototype Software EVK1-CT-2000-00053, report D13

Røstum, J, Kowalski M, Hulance, J: User manual CARE – W Rehab manager. EVK1-CT-2000-00053 report D14, Jan 2004.

Sægrov S, Schilling W: Computer Aided Rehabilitation of Water networks, 10. International Conference on Urban Drainage, Copenhagen 23 August 2005

Volta M, Poulton M, Herrero M, Røstum J, Sægrov S: Testing and validation of CARE-W Rehab manager. EVK1-CT-2000-00053 report D16, Jan 2004

Appendix 1

Definitions related to water network rehabilitation

Definitions

In this section, definitions are presented for a number of key expressions which are generally used in rehabilitation and whose precise meaning must be made clear:

- Active leakage control: A formal regime of regular (periodic) or continuous monitoring of the network that detects and repairs unreported bursts; includes regular surveying (sounding, waste metering) and/or leakage monitoring. Does not include repair of reported bursts.
- Complaints: Any customer complaint expressed to the services either personally (verbal or written), by telephone, fax, post, e-mail or any other written format.
- Inspection: Formal, probably written, procedure that is subsequently recorded in a form that enables the undertaking to assess the serviceability of the assets and to take corrective action as appropriate.
- Interruption: Unplanned (even if notified) or un-notified water supply interruption to customers with a duration (measured to full restoration of supply) of more than 12 hours, caused by bursts or failures in the water supply system and the subsequent repair/renewal measures. Includes those planned interruptions that exceed the notified period.
- Mains failure: Detected water leaks of transmission and/or distribution mains necessitating repair / renewal measures. Included are failures of mains, defective pipe connections, valves and fittings, caused by:
  - defective materials, design, construction or operational-related defects, in pipes, joints, valves and other fittings;
  - corrosion of materials, externally or internally, primarily but not exclusively ferrous materials;
  - external mechanical damage e.g. due to excavation, including third party damage;
  - earth movements related to effects of frost, dry periods, heavy traffic, earthquakes, floods, and others.
- Rehabilitation: Any physical intervention that extends the life of the system and involves changing their condition or specification.
- Refurbishment: All methods for restoring the existing assets in order to achieve the requisite performance.
- Relining: The removal of all deposits from inside an existing pipeline, followed by the in situ application of a non-structural lining to provide corrosion protection, such as cement or epoxy mortar (relining is sometimes referred to as scraping and lining, renovation or reconditioning).

- Replacement: Substitution of a new facility for an existing one where the latter is no longer used for its former objective. Renewal is a particular form of replacement in which the function of the new facility is the same as that of the existing. In practice this usually means that is of the same nominal diameter (for pipelines), power (for pumping systems), etc. In the case of pipelines, replacement includes the provision of a structural liner (sliplining). The new pipeline may or may not have the same carrying capacity as the existing pipeline.
- Restrictions to water service: Limitations to the use of water, as imposed by the water undertaking, such as hosepipe or sprinkler bans.
- Service connection: The authorised pipe connecting the main to the measurement point or to the customer stop-valve, as applicable. Where several registered customers or individually occupied premises share a physical connection or tapping off the main, e.g. apartment buildings, this will still be regarded as the one connection for the purposes of the applicable PI, irrespective of the configuration and number of customers or premises.
- Service connection failure: Same definition as mains failure applied to service connection.
- Water losses: The difference between system input volume and authorised consumption. Water losses can be considered as a total volume for the whole system, or for partial systems such as raw water mains, transmission or distribution. In each case the components of the calculation would be adjusted accordingly. Water losses consist of real losses and apparent losses.
- Real losses: Physical water losses from the pressurised system, up to the point of measurement of customer use. The annual volume lost through all types of leaks, bursts and overflows depends on frequencies, flow rates, and average duration of individual leaks. Note: Although physical losses after the point of customer flow measurement are excluded from the assessment of real losses, they are often significant (particularly where customers are unmetered) and worthy of attention for demand management purposes.

Appendix 2

CARE-W rehab performance indicators

PROPOSAL OF REHAB PERFORMANCE INDICATORS

Note: The IWA codes for PI and PI variables are kept in this Appendix to allow for an easier cross-reference. See [2] for variables identification when necessary.

Operational indicators

INDICATOR (unit)	CONCEPT / Processing rule
REHABILITATION	
Op15 - Mains rehabilitation *(%/year)*	Length of transmission and distribution mains rehabilitated during the year / total mains length x 100 *Op15 = D18/C6 x 100 or Op15 = Op16 + Op17*
• *Op16 - mains relining*	Length of mains relined during the year / total mains length x 100 *Op16 = D19/C6 x 100*
• *Op17 - replaced or renewed mains*	Length of mains replaced or renewed during the year / total mains length x 100 *Op17 = D20/C6 x 100*
• *Op18 - replaced or renewed valves* *(%/year)*	Number of mains valves replaced or renewed during the year / total number of mains valves x 100 *Op18 = D21/C29 x 100*
Op19 - Service connection rehabilitation *(%/year)*	Number of service connections replaced or renewed during the year / total number of service connections x 100 *Op19 = D22/C32 x 100*
FAILURES AND REPAIRS	
Op26 - Mains failures[1] *(No./100 km/year)*	Number of mains failures during the year, including failures of pipes, valves, fittings and service connection insertion point failures / total mains length x 100 *Op26 = D25/C6 x 100* *If mains failures are to be used for regulating objectives, the use of a complementary indicator, similar to Op26 but excluding failures by third parties is advisable, as they are not a direct fault of the water undertaking. Number should exclude repairs under active leakage control.*
Op26a - pipe failures *(No./100 km/year)*	Number of pipe failures during the year / total mains length x 100 *Op26a = D25a/C6 x 100*
Op26b - joint failures *(No./100 km/year)*	Number of joint failures during the year / total mains length x 100 *Op26b = D25b/C6 x 100*
Op26c - valve failures *(No./100 km/year)*	Number of valve failures during the year / total mains length x 100 *Op26c = D25c/C6 x 100*
Op26d - service connection insertion point failures *(No./100 km/year)*	Number of failures that occur in the insertion point of the service connection during the year / total mains length x 100 *Op26d = D25d/C6 x 1000*
Op26e - Critical mains failures *(No./100 km/year)*	Number of mains failures in sensitive areas during the year / total mains length in sensitive areas x 100 *Op26e = D25f/C6a x 100* *Sensitive areas can be defined as urban zones where serious human consequences (for example hospitals, dialysis patients, etc), severe damages (landslide, flooding) or severe disturbances (traffic interruptions) may occur due to water main bursts.*
Op27 - Service connection failures *(No./1000 connections/year)*	Number of service connection failures during the year / number of service connections x 1000 *Op27 = D26/C32 x 1000* *If service connection failures are to be used for regulating objectives, the use of a complementary indicator, similar to Op27 but excluding failures by third parties is*

[1] This definition differs from the present IWA's one. In the latter it is not clear whether the service connection insertion point failures shall be accounted for.

INDICATOR *(unit)*	CONCEPT Processing rule
	advisable, as they are not a direct fault of the water undertaking. Number should exclude repairs under active leakage control.
Op28 - Hydrant failures *(No./1000 hydrants/year)*	Number of hydrant failures during the year / total number of hydrants x 1000 *Op28 = D27/C31 x 1000* *If hydrant failures are to be used for regulating objectives, the use of a complementary indicator, similar to Op28 but excluding failures by third parties is advisable, as they are not a direct fault of the water undertaking. Number should exclude repairs under active leakage control.*
Op29 - Power failures *(hours/pumping station/year)*	$\Sigma_{\text{(for all pumping stations)}}$ (number of hours during the year each pumping station is out of service or is reliant on standby power generation due to power supply interruptions)/ total number of pumping stations *Op29 = D28/C4*
Op5 - Active leakage control repairs *(No./100 km/year)*	Number of main leaks detected and repaired due to active leakage control / total mains length x 100 *Op5 = D9/C6 x 100*
WATER LOSSES	(Water balance definitions to be adopted according to [2])
Op22 - Water losses per connection *(m^3/connection/year)*	Water losses / number of service connections *Op22 = A20/C32* *If service connections density < 20/km of mains (e.g. transmission networks), then this indicator should be expressed in m^3/ km of water mains / year*
• *Op24 - real losses* per connection *(l/connection/day when system is pressurised)*	Real losses x 1000 / (number of service connections x 365 x T/100) (T = % of year system is pressurised) *Op24 = A24 x 1000/(C32x365xD29/365/24)* *If service connections density < 20/km of mains (e.g. transmission networks), then this indicator should be expressed in l / km of water mains / day*
Op25 - Infrastructure leakage index *(-)*	Real losses (*Op24*)/ technical achievable low-level annual real losses (when system is pressurised) *Op25 = Op24/(18xC6/C32+0.7+0.025xC33)xD31* *The technical achievable low-level annual real losses are equal to the "best estimate" of so called Unavoidable Annual Real Losses, UARL, which can be calculated with the equation derived by the Water Losses Task Force (see AQUA Dec. 1999 and IWA Blue Pages "Losses from water supply systems"):* *UARL (litres/service connection/day)= (18 x Lm/Nc + 0,7 + 0,025 Lp) x P* *This equation, based on empirical results of international investigations, recognises separate influences on real losses from:* - *Length of mains Lm in km (C6)* - *Number of service connections Nc (C32)* - *Average length of service connections Lp in m (C33)* - *Average operating pressure P in m (D31)* *Well-managed systems are expected to have low values of this Infrastructure leakage index – close to 1.0 – while systems with infrastructure management deficiencies will present higher values.*

Quality of service indicators

INDICATOR *(unit)*	CONCEPT *Processing rule*
SERVICE	
QS9 - Pressure of supply adequacy *(%)*	Number of delivery points that receive and are likely to receive pressure equal to or above the guaranteed or declared target level at the peak demand hour (but not when demand is abnormal) / number service connections x 100 *QS9 = D30/C32 x 100*
QS11 - Water interruptions[2] *(%)*	Σ (Population subject to a water interruption x duration of the interruption in hours) / (population served x 24 x 365) x 100 *QS11 = D32/(F1x24x365) x 100* *Since, for many water undertakings, the information required for this indicator is neither available nor feasible to be collected in a near future, QS12 is alternatively proposed.*
QS12 - Interruptions per connection *(No./1000 connections)*	Number of interruptions / number of service connections x 1000 *QS12 = D33/C32 x 1000* *This indicator should only be used if QS cannot be calculated.*
• *QS12a - critical interruptions per connection* *(No./1000 connections)*	Number of supply interruptions to sensitive service connections during the year / number of sensitive service connections x 1000 *QS12a = D33a/C32a x 1000* *Sensitive service connections can be defined as connections where serious human consequences may occur due to supply interruptions (for example hospitals, dialysis patients, etc).*
QS13 - Population experiencing restrictions to water service[3] *(%)*	Σ (Population affected by restrictions to water service x duration of the restrictions to water service in hours) / (total population served x 24 x x 365) x 100 *QS13 = D34/(F1x24x365) x 100*
QS14 - Days with restrictions to water service *(%)*	Total number of days with restrictions to water service during the year / 365 x 100 *QS14 = D35/365 x 100* *This indicator should only be used if QS13 cannot be calculated.*
QS15 - Quality of supplied water *(%)*	Total number of treated water tests complying with the applicable standards or legislation during the year / total number of tests of treated water performed during the year x 100 *QS15 = (D51+D52+D53)/D41 x 100*
• *QS16 - aesthetic* tests compliance *(%)*	Number of aesthetic tests of treated water complying with the applicable standards or legislation during the year / total number of aesthetic tests treated water performed during the year x 100 *QS16 = D51/D42 x 100*
• *QS16a - water taste* tests compliance *(%)*	Number of water taste tests of treated water complying with the applicable standards or legislation during the year / total number of water taste tests of treated water performed during the year x 100 *QS16a = D51a/D42a x 100*
• *QS16b - water colour* tests compliance *(%)*	Number of water colour tests of treated water complying with the applicable standards or legislation during the year / total number of water colour tests of treated water performed during the year x 100 *QS16 b= D51b/D42b x 100*
• *QS17 - microbiological* tests compliance *(%)*	Number of microbiological tests of treated water complying with the applicable standards or legislation during the year / total number of microbiological tests of treated water performed during the year x 100 *QS17 = D52/D43 x 100*
• *QS18 - physical-chemical* tests compliance *(%)*	Number of physical-chemical tests of treated water complying with the applicable standards or legislation during the year / total number of physical-chemical tests of treated water performed during the year x 100 *QS18 = D53/D44 x 100*

[2] Indicators *QS11* and *QS12* shall be used in alternative.
[3] Indicators *QS13* and *QS14* shall be used in alternative.

INDICATOR (unit)	CONCEPT Processing rule
CUSTOMER COMPLAINTS	
QS22 - Service complaints per connection *(No. complaints/1000 connections/year)*	Number of service complaints during the year / number of service connections x 1000 *QS22 = F11/C32 x 1000*
• *QS23 - pressure complaints* *(%)*	Number of pressure complaints during the year / number of service complaints during the year x 100 *QS23 = F12/F11 x 100*
• *QS24 - continuity complaints* *(%)*	Number of continuity complaints during the year / number of service complaints during the year x 100 *QS24 = F13/F11 x 100* *This indicator refers to medium- or long-term supply constraints, due to insufficient water quantity or quality at source, insufficient system capacity or on-going works.*
• *QS25 - water quality complaints* *(%)*	Number of water quality complaints during the year / number of service complaints during the year x 100 *QS25 = F14/F11 x 100*
• *QS25a - water taste complains* *(%)*	Number of water taste complaints during the year / number of service complaints during the year x 100 *QS25a = F14a/F11 x 100*
• *QS25b - water colour complains* *(%)*	Number of water colour complaints during the year / number of service complaints during the year x 100 *QS25b = F14b/F11 x 100*
• *QS26 - interruptions complains* *(%)*	Number of complaints due to supply interruptions during the year / number of service complaints during the year x 100 *QS26 = F15/F11 x 100* *This indicator refers to short-term supply interruptions, due to accidental system failures or repair works.*
• *QS26a - critical interruptions complains* *(%)*	Number of complaints due to supply interruptions to sensitive service connections during the year / number of service complaints during the year x 100 *QS26a= F15a/F11 x 100* *Sensitive service connections can be defined as connections where serious human consequences may occur due to supply interruptions (for example hospitals, dialysis patients, etc).* *This indicator refers to short-term supply interruptions, due to accidental system failures or repair works.*

Financial Indicators

INDICATOR (unit)	CONCEPT Processing rule
ANNUAL COSTS	(Refer to [2] for definitions)
Fi1 - Unit total costs *(€/m³)*	(Annual running costs + annual capital costs) / authorised consumption (including exported water) *Fi1 = (G2+G3)/A19*
• *Fi2 - unit running costs* *(€/m³)*	Annual running costs / authorised consumption (including exported water) *Fi2 = G2/A19*
ANNUAL INVESTMENT FOR NETWORK MAINS	
Fi18a - Unit investment for network mains *(€/m³)*	Annual cost of investments (expenditures for network mains) / authorised consumption (including exported water) *Fi18a = G26/A19* *The annual values of this ratio can be misleading. A multi-annual analysis must be adopted.*

INDICATOR (unit)	CONCEPT Processing rule
• *Fi19a - annual investments for new and upgrading mains*	Cost of investments for new mains (or upgrading of existing ones) / total cost of the investments x 100 *Fi19a = G27/G26 x 100* *The annual values of this ratio can be misleading. A multi-annual analysis must be adopted.*
• *Fi20a - annual investments for mains replacement* (%)	Cost of investments for the replacement of existing mains/ cost of the investments x 100 *Fi20a = G28/G26 x 100* *The annual values of this ratio can be misleading. A multi-annual analysis must be adopted.*
TARIFFS	
Fi21 - Average water charges for direct consumption (€/m^3)	Annual water sales revenue from residential, commercial, industrial, public, institutional and other customers (exported water excluded; public water taxes excluded) / (total annual authorised - exported water) *Fi21 = G30/(A19-A9)*
Fi22 - Average water charges for exported water (€/m^3)	Annual water sales revenue from exported water (excluding public water taxes) / exported water *Fi22 = G31/(A6+A9)*

Water resources indicators

INDICATOR (unit)	CONCEPT Processing rule
WR1 - Inefficiency of use of water resources (%)	Real losses / system input volume x 100 *WR1 = A24/(A4+A5+A8) x 100* *This indicator must not be used as a measure of efficiency of management of the transmission and/or the distribution system.*
WR2 - Annual water resources availability ratio (%)	System input volume / total annual abstraction capacity and imported water allowance x 100 *WR2 = (A19+A20)/(A1+A2) x 100 or WR2 = (A7+ A8)/(A1+A2) x 100* *A value of 100% for this indicator means that all available resources are being used. Although this indicator is sometimes difficult to assess, and is not easily auditable, its used is encouraged as a management tool, particularly in rapid growing areas or areas subject to scarcity problems. Each water undertaking should estimate the annual abstraction capacity and imported water allowance taking into account water quality requirements and according to its own guaranteed schemes, drought management and operation procedures. This indicator is not suitable for comparisons, unless the same underlying assumptions are adopted to assess A1.*

Physical indicators

INDICATOR (unit)	CONCEPT Processing rule
Ph3 - Transmission and distribution storage capacity (days)	Total capacity of transmission and distribution storage tanks (private storage tanks excluded) / [authorised consumption (including exported water) + water losses] x 365 *Ph3=C2/(A19+A20) x 365* *In case of bulk supply systems, if the delivery point is a storage tank, its capacity can be accounted for, even though it is neither owned nor operator by the water undertaking. The interpretation of the values of this indicator shall take into account the seasonal, monthly, daily and hourly peak factors.*
Ph7 - Valve density (No./km)	Number of isolating valves / total distribution mains length *Ph7 = C30/C6*

Appendix 3

CARE-W rehab utility information

PROPOSAL OF REHAB UTILITY INFORMATION

Note: The IWA codes for PI variables [2] are kept in this Appendix to allow for an easier cross-reference.

PHYSICAL ASSETS DATA - DISTRIBUTION NETWORK

TRANSMISSION AND DISTRIBUTION NETWORK

C6 - MAINS LENGTH (IWA PI variable)

UNIT OF EXPRESSION: km	REFERENCE DATE: *[dd.mm]*.yy	TYPE: Real

DEFINITION:

Total transmission and distribution mains length (service connections not included).

COMMENT:

Mains that are not yet in use or have been put out of service on a permanent basis shall not be accounted for.

C6a - MAINS LENGTH IN SENSITIVE AREAS (subset of the IWA PI variable C6)

UNIT OF EXPRESSION: km	REFERENCE DATE: *[dd.mm]*.yy	TYPE: Real

DEFINITION:

Total transmission and distribution mains length (service connections not included).

COMMENT:

Mains that are not yet in use or have been put out of service on a permanent basis shall not be accounted for.

Sensitive areas can be defined as urban zones where serious human consequences (for example hospitals, dialysis patients, etc), severe damages (landslide, flooding) or severe disturbances (traffic interruptions) may occur due to water main bursts.

C6b - NETWORK EXTENSION (subset of the IWA PI variable C6)

UNIT OF EXPRESSION: km/year	REFERENCE DATE: *[dd.mm]*.yy	TYPE: Real

DEFINITION:

Average annual extension of transmission and distribution mains length over the last 5 years (service connections excluded)

COMMENT:

C15 - MAINS WITH < 100/110 mm DIAMETER (IWA PI variable)

UNIT OF EXPRESSION: km	REFERENCE DATE: *[dd.mm]*.yy	TYPE: Real

DEFINITION:

Total length of transmission and distribution mains with internal/external diameter $\leq$ 100/110 mm (Service connections not included).

COMMENT:

It is recommended that, whenever feasible, this variable is split into sub-variables according to the pipe material, to allow for a better interpretation of the pipe failure information.

C16 - MAINS WITH 100/110 < DIAMETER ≤ 300/315 mm (IWA PI variable)		
UNIT OF EXPRESSION: km	REFERENCE DATE: *[dd.mm]*.yy	TYPE: Real
DEFINITION: Total length of transmission and distribution mains with internal/external diameter > 100/110 mm and ≤ 300/315 mm (Service connections not included).		
COMMENT: It is recommended that, whenever feasible, this variable is split into sub-variables according to the pipe material, to allow for a better interpretation of the pipe failure information.		

C17 - MAINS WITH DIAMETER > 300/315 mm (IWA PI variable)		
UNIT OF EXPRESSION: km	REFERENCE DATE: *[dd.mm]*.yy	TYPE: Real
DEFINITION: Total length of transmission and distribution mains with internal/external diameter > 300/315 mm (Service connections not included).		
COMMENT: It is recommended that, whenever feasible, this variable is split into sub-variables according to the pipe material, to allow for a better interpretation of the pipe failure information.		

C7 - CAST SPUN AND GREY IRON MAINS (IWA PI variable)		
UNIT OF EXPRESSION: km	REFERENCE DATE: *[dd.mm]*.yy	TYPE: Real
DEFINITION: Total length of cast iron transmission and distribution mains (Service connections not included).		
COMMENT:		

C8 - DUCTILE IRON MAINS (IWA PI variable)		
UNIT OF EXPRESSION: km	REFERENCE DATE: *[dd.mm]*.yy	TYPE: Real
DEFINITION: Total length of ductile iron transmission and distribution mains (Service connections not included).		
COMMENT:		

C9 - STEEL MAINS (IWA PI variable)		
UNIT OF EXPRESSION: km	REFERENCE DATE: *[dd.mm]*.yy	TYPE: Real
DEFINITION: Total length of steel transmission and distribution mains (Service connections not included).		
COMMENT:		

C10 - ASBESTOS CEMENT MAINS (IWA PI variable)		
UNIT OF EXPRESSION: km	REFERENCE DATE: *[dd.mm]*.yy	TYPE: Real
DEFINITION: Total length of asbestos cement transmission and distribution mains (Service connections not included).		
COMMENT:		

C11 - POLYETHYLENE MAINS (IWA PI variable)		
UNIT OF EXPRESSION: km	REFERENCE DATE: *[dd.mm]*.yy	TYPE: Real
DEFINITION: Total length of polyethylene transmission and distribution mains (Service connections not included).		
COMMENT:		

C12 - POLYVINYL CHLORINE MAINS (IWA PI variable)		
UNIT OF EXPRESSION: km	REFERENCE DATE: *[dd.mm]*.yy	TYPE: Real
DEFINITION: Total length of polyvinyl chlorine transmission and distribution mains (Service connections not included).		
COMMENT:		

C13 - CONCRETE MAINS (IWA PI variable)		
UNIT OF EXPRESSION: km	REFERENCE DATE: *[dd.mm]*.yy	TYPE: Real
DEFINITION: Total length of concrete transmission and distribution mains (Service connections not included).		
COMMENT:		

C14 - OTHER MATERIAL MAINS (IWA PI variable)		
UNIT OF EXPRESSION: km	REFERENCE DATE: *[dd.mm]*.yy	TYPE: Real
DEFINITION: Total length of transmission and distribution mains made of other material than cast iron, ductile iron, steel, asbestos cement, polyethylene, polyvinyl chlorine and concrete (Service connections not included).		
COMMENT: If appropriate, this variable can be split into sub-variables.		

C34 - MAINS WITH INTERNAL PROTECTION		
UNIT OF EXPRESSION: km	REFERENCE DATE: *[dd.mm]*.yy	TYPE: Real
DEFINITION: Total length of transmission and distribution mains with internal protection (Service connections not included).		
COMMENT: If appropriate, this variable can be split into sub-variables.		

C35 - MAINS WITH EXTERNAL PROTECTION)		
UNIT OF EXPRESSION: km	REFERENCE DATE: *[dd.mm]*.yy	TYPE: Real
DEFINITION: Total length of transmission and distribution mains with external protection (Service connections not included).		
COMMENT: If appropriate, this variable can be split into sub-variables.		

C36 - MAINS WITH CATHODIC PROTECTION

UNIT OF EXPRESSION: km	REFERENCE DATE: *[dd.mm]*.yy	TYPE: Real

DEFINITION:

Total length of transmission and distribution mains with cathodic protection (Service connections not included).

COMMENT:

If appropriate, this variable can be split into sub-variables.

C37 - AVERAGE MAINS AGE

UNIT OF EXPRESSION: years	REFERENCE DATE: *[dd.mm]*.yy	TYPE: Real

DEFINITION:

Average mains age for the global supply system based on the age of each mains and its length

COMMENT:

It is recommended that, whenever feasible, this variable is split into sub-variables according to the pipe material, to allow for a better interpretation of the pipe failure information.

It is advisable that the water undertaking defines a clear procedure to establish the pipe age of partially renewed or rehabilitated pipes, to allow for a correct interpretation of the results.

C18a - MAINS LAID BETWEEN 2006 AND 2010

UNIT OF EXPRESSION: km	REFERENCE DATE: *[dd.mm]*.yy	TYPE: Real

DEFINITION:

Total length of transmission and distribution mains laid between 2006 and 2010 (Service connections not included).

COMMENT:

It is recommended that, whenever feasible, this variable is split into sub-variables according to the pipe material, to allow for a better interpretation of the pipe failure information.

It is advisable that the water undertaking defines a clear procedure to establish the pipe age of partially renewed or rehabilitated pipes, to allow for a correct interpretation of the results.

C18b - MAINS LAID BETWEEN 2001 AND 2005

UNIT OF EXPRESSION: km	REFERENCE DATE: *[dd.mm]*.yy	TYPE: Real

DEFINITION:

Total length of transmission and distribution mains laid between 2001 and 2005 (Service connections not included).

COMMENT:

It is recommended that, whenever feasible, this variable is split into sub-variables according to the pipe material, to allow for a better interpretation of the pipe failure information.

It is advisable that the water undertaking defines a clear procedure to establish the pipe age of partially renewed or rehabilitated pipes, to allow for a correct interpretation of the results.

C18c - MAINS LAID BETWEEN 1996 AND 2000

UNIT OF EXPRESSION: km	REFERENCE DATE: *[dd.mm]*.yy	TYPE: Real

DEFINITION:

Total length of transmission and distribution mains laid between 1996 and 2000 (Service connections not included).

COMMENT:

It is recommended that, whenever feasible, this variable is split into sub-variables according to the pipe material, to allow for a better interpretation of the pipe failure information.

It is advisable that the water undertaking defines a clear procedure to establish the pipe age of partially renewed or rehabilitated pipes, to allow for a correct interpretation of the results.

C18d - MAINS LAID BETWEEN 1991 AND 1995

UNIT OF EXPRESSION: km	REFERENCE DATE: *[dd.mm]*.yy	TYPE: Real

DEFINITION:

Total length of transmission and distribution mains laid between 1991 and 1995 (Service connections not included).

COMMENT:

It is recommended that, whenever feasible, this variable is split into sub-variables according to the pipe material, to allow for a better interpretation of the pipe failure information.

It is advisable that the water undertaking defines a clear procedure to establish the pipe age of partially renewed or rehabilitated pipes, to allow for a correct interpretation of the results.

C18e - MAINS LAID BETWEEN 1986 AND 1990

UNIT OF EXPRESSION: km	REFERENCE DATE: *[dd.mm]*.yy	TYPE: Real

DEFINITION:

Total length of transmission and distribution mains laid between 1986 and 1990 (Service connections not included).

COMMENT:

It is recommended that, whenever feasible, this variable is split into sub-variables according to the pipe material, to allow for a better interpretation of the pipe failure information.

It is advisable that the water undertaking defines a clear procedure to establish the pipe age of partially renewed or rehabilitated pipes, to allow for a correct interpretation of the results.

C18f - MAINS LAID BETWEEN 1981 AND 1985

UNIT OF EXPRESSION: km	REFERENCE DATE: *[dd.mm]*.yy	TYPE: Real

DEFINITION:

Total length of transmission and distribution mains laid between 1981 and 1985 (Service connections not included).

COMMENT:

It is recommended that, whenever feasible, this variable is split into sub-variables according to the pipe material, to allow for a better interpretation of the pipe failure information.

It is advisable that the water undertaking defines a clear procedure to establish the pipe age of partially renewed or rehabilitated pipes, to allow for a correct interpretation of the results.

C18g - MAINS LAID BETWEEN 1976 AND 1980

UNIT OF EXPRESSION: km	REFERENCE DATE: *[dd.mm]*.yy	TYPE: Real

DEFINITION:

Total length of transmission and distribution mains laid between 1976 and 1980 (Service connections not included).

COMMENT:

It is recommended that, whenever feasible, this variable is split into sub-variables according to the pipe material, to allow for a better interpretation of the pipe failure information.

It is advisable that the water undertaking defines a clear procedure to establish the pipe age of partially renewed or rehabilitated pipes, to allow for a correct interpretation of the results.

C19a - MAINS LAID BETWEEN 1971 AND 1975

UNIT OF EXPRESSION: km	REFERENCE DATE: *[dd.mm]*.yy	TYPE: Real

DEFINITION:

Total length of transmission and distribution mains laid between 1971 and 1975 (Service connections not included).

COMMENT:

It is recommended that, whenever feasible, this variable is split into sub-variables according to the pipe material, to allow for a better interpretation of the pipe failure information.

It is advisable that the water undertaking defines a clear procedure to establish the pipe age of partially renewed or rehabilitated pipes, to allow for a correct interpretation of the results.

C19b - MAINS LAID BETWEEN 1966 AND 1970

UNIT OF EXPRESSION: km	REFERENCE DATE: *[dd.mm]*.yy	TYPE: Real

DEFINITION:

Total length of transmission and distribution mains laid between 1976 and 1970 (Service connections not included).

COMMENT:

It is recommended that, whenever feasible, this variable is split into sub-variables according to the pipe material, to allow for a better interpretation of the pipe failure information.

It is advisable that the water undertaking defines a clear procedure to establish the pipe age of partially renewed or rehabilitated pipes, to allow for a correct interpretation of the results.

C19c - MAINS LAID BETWEEN 1961 AND 1965

UNIT OF EXPRESSION: km	REFERENCE DATE: *[dd.mm]*.yy	TYPE: Real

DEFINITION:

Total length of transmission and distribution mains laid between 1961 and 1965 (Service connections not included).

COMMENT:

It is recommended that, whenever feasible, this variable is split into sub-variables according to the pipe material, to allow for a better interpretation of the pipe failure information.

It is advisable that the water undertaking defines a clear procedure to establish the pipe age of partially renewed or rehabilitated pipes, to allow for a correct interpretation of the results.

C19d - MAINS LAID BETWEEN 1956 AND 1960		
UNIT OF EXPRESSION: km	REFERENCE DATE: *[dd.mm]*.yy	TYPE: Real
DEFINITION: Total length of transmission and distribution mains laid between 1956 and 1960 (Service connections not included).		
COMMENT: It is recommended that, whenever feasible, this variable is split into sub-variables according to the pipe material, to allow for a better interpretation of the pipe failure information. It is advisable that the water undertaking defines a clear procedure to establish the pipe age of partially renewed or rehabilitated pipes, to allow for a correct interpretation of the results.		

C19d - MAINS LAID BETWEEN 1951 AND 1955		
UNIT OF EXPRESSION: km	REFERENCE DATE: *[dd.mm]*.yy	TYPE: Real
DEFINITION: Total length of transmission and distribution mains laid between 1951 and 1955 (Service connections not included).		
COMMENT: It is recommended that, whenever feasible, this variable is split into sub-variables according to the pipe material, to allow for a better interpretation of the pipe failure information. It is advisable that the water undertaking defines a clear procedure to establish the pipe age of partially renewed or rehabilitated pipes, to allow for a correct interpretation of the results.		

C20a - MAINS LAID BETWEEN 1946 AND 1950		
UNIT OF EXPRESSION: km	REFERENCE DATE: *[dd.mm]*.yy	TYPE: Real
DEFINITION: Total length of transmission and distribution mains laid between 1946 and 1950 (Service connections not included).		
COMMENT: It is recommended that, whenever feasible, this variable is split into sub-variables according to the pipe material, to allow for a better interpretation of the pipe failure information. It is advisable that the water undertaking defines a clear procedure to establish the pipe age of partially renewed or rehabilitated pipes, to allow for a correct interpretation of the results.		

C20b - MAINS LAID BETWEEN 1941 AND 1945		
UNIT OF EXPRESSION: km	REFERENCE DATE: *[dd.mm]*.yy	TYPE: Real
DEFINITION: Total length of transmission and distribution mains laid between 1941 and 1945 (Service connections not included).		
COMMENT: It is recommended that, whenever feasible, this variable is split into sub-variables according to the pipe material, to allow for a better interpretation of the pipe failure information. It is advisable that the water undertaking defines a clear procedure to establish the pipe age of partially renewed or rehabilitated pipes, to allow for a correct interpretation of the results.		

C20c - MAINS LAID BETWEEN 1936 AND 1940		
UNIT OF EXPRESSION: km	REFERENCE DATE: *[dd.mm]*.yy	TYPE: Real
DEFINITION: Total length of transmission and distribution mains laid between 1936 and 1940 (Service connections not included).		
COMMENT: It is recommended that, whenever feasible, this variable is split into sub-variables according to the pipe material, to allow for a better interpretation of the pipe failure information. It is advisable that the water undertaking defines a clear procedure to establish the pipe age of partially renewed or rehabilitated pipes, to allow for a correct interpretation of the results.		

C20d - MAINS LAID BETWEEN 1931 AND 1935		
UNIT OF EXPRESSION: km	REFERENCE DATE: *[dd.mm]*.yy	TYPE: Real
DEFINITION: Total length of transmission and distribution mains laid between 1931 and 1935 (Service connections not included).		
COMMENT: It is recommended that, whenever feasible, this variable is split into sub-variables according to the pipe material, to allow for a better interpretation of the pipe failure information. It is advisable that the water undertaking defines a clear procedure to establish the pipe age of partially renewed or rehabilitated pipes, to allow for a correct interpretation of the results.		

C20e - MAINS LAID BETWEEN 1926 AND 1930		
UNIT OF EXPRESSION: km	REFERENCE DATE: *[dd.mm]*.yy	TYPE: Real
DEFINITION: Total length of transmission and distribution mains laid between 1926 and 1930 (Service connections not included).		
COMMENT: It is recommended that, whenever feasible, this variable is split into sub-variables according to the pipe material, to allow for a better interpretation of the pipe failure information. It is advisable that the water undertaking defines a clear procedure to establish the pipe age of partially renewed or rehabilitated pipes, to allow for a correct interpretation of the results.		

C21a - MAINS LAID BETWEEN 1921 AND 1925		
UNIT OF EXPRESSION: km	REFERENCE DATE: *[dd.mm]*.yy	TYPE: Real
DEFINITION: Total length of transmission and distribution mains laid between 1921 and 1925 (Service connections not included).		
COMMENT: It is recommended that, whenever feasible, this variable is split into sub-variables according to the pipe material, to allow for a better interpretation of the pipe failure information. It is advisable that the water undertaking defines a clear procedure to establish the pipe age of partially renewed or rehabilitated pipes, to allow for a correct interpretation of the results.		

C21b - MAINS LAID BETWEEN 1916 AND 1920		
UNIT OF EXPRESSION: km	REFERENCE DATE: *[dd.mm]*.yy	TYPE: Real
DEFINITION: Total length of transmission and distribution mains laid between 1916 and 1920 (Service connections not included).		
COMMENT: It is recommended that, whenever feasible, this variable is split into sub-variables according to the pipe material, to allow for a better interpretation of the pipe failure information. It is advisable that the water undertaking defines a clear procedure to establish the pipe age of partially renewed or rehabilitated pipes, to allow for a correct interpretation of the results.		

C21c - MAINS LAID BETWEEN 1911 AND 1915		
UNIT OF EXPRESSION: km	REFERENCE DATE: *[dd.mm]*.yy	TYPE: Real
DEFINITION: Total length of transmission and distribution mains laid between 1921 and 1925 (Service connections not included).		
COMMENT: It is recommended that, whenever feasible, this variable is split into sub-variables according to the pipe material, to allow for a better interpretation of the pipe failure information. It is advisable that the water undertaking defines a clear procedure to establish the pipe age of partially renewed or rehabilitated pipes, to allow for a correct interpretation of the results.		

C21d - MAINS LAID BETWEEN 1906 AND 1910		
UNIT OF EXPRESSION: km	REFERENCE DATE: *[dd.mm]*.yy	TYPE: Real
DEFINITION: Total length of transmission and distribution mains laid between 1906 and 1910 (Service connections not included).		
COMMENT: It is recommended that, whenever feasible, this variable is split into sub-variables according to the pipe material, to allow for a better interpretation of the pipe failure information. It is advisable that the water undertaking defines a clear procedure to establish the pipe age of partially renewed or rehabilitated pipes, to allow for a correct interpretation of the results.		

C21e - MAINS LAID BETWEEN 1901 AND 1905		
UNIT OF EXPRESSION: km	REFERENCE DATE: *[dd.mm]*.yy	TYPE: Real
DEFINITION: Total length of transmission and distribution mains laid between 1901 and 1905 (Service connections not included).		
COMMENT: It is recommended that, whenever feasible, this variable is split into sub-variables according to the pipe material, to allow for a better interpretation of the pipe failure information. It is advisable that the water undertaking defines a clear procedure to establish the pipe age of partially renewed or rehabilitated pipes, to allow for a correct interpretation of the results.		

C21f - MAINS LAID BEFORE 1900		
UNIT OF EXPRESSION: km	REFERENCE DATE: *[dd.mm]*.yy	TYPE: Real
DEFINITION: Total length of transmission and distribution mains laid before 1900 (Service connections not included).		
COMMENT: It is recommended that, whenever feasible, this variable is split into sub-variables according to the pipe material, to allow for a better interpretation of the pipe failure information. It is advisable that the water undertaking defines a clear procedure to establish the pipe age of partially renewed or rehabilitated pipes, to allow for a correct interpretation of the results.		

C38 - MAINS LOCATION UNDER FLEXIBLE ROADWAY)		
UNIT OF EXPRESSION: km	REFERENCE DATE: *[dd.mm]*.yy	TYPE: Real
DEFINITION: Mains location under flexible roadway		
COMMENT:		

C39 - MAINS LOCATION UNDER RIGID ROADWAY		
UNIT OF EXPRESSION: km	REFERENCE DATE: *[dd.mm]*.yy	TYPE: Real
DEFINITION: Mains location under rigid roadway		
COMMENT:		

C40 - MAINS LOCATION UNDER SIDEWALK		
UNIT OF EXPRESSION: km	REFERENCE DATE: *[dd.mm]*.yy	TYPE: Real
DEFINITION: Mains location under sidewalk		
COMMENT:		

C41 - MAINS LOCATION UNDER GREEN AREAS		
UNIT OF EXPRESSION: km	REFERENCE DATE: *[dd.mm]*.yy	TYPE: Real
DEFINITION: Mains location under green areas		
COMMENT:		

C42 - MAINS INSTALLATION DEPTH		
UNIT OF EXPRESSION: m	REFERENCE DATE: *[dd.mm]*.yy	TYPE: Real
DEFINITION: Average mains installation depth from the pavement to the external top of the pipe		
COMMENT:		

C43 – MAINS WITH APPROPRIATE BEDDING		
UNIT OF EXPRESSION: km	REFERENCE DATE: *[dd.mm]*.yy	TYPE: Alpha-numeric
DEFINITION: Length of mains with appropriate bedding compactation and side support		
COMMENT:		

C43a – MAINS WITH INADEQUATE BEDDING		
UNIT OF EXPRESSION: km	REFERENCE DATE: *[dd.mm]*.yy	TYPE: Alpha-numeric
DEFINITION: Length of mains with inadequate bedding compactation or side support		
COMMENT:		

C44 – MAINS WITH VERY AGGRESSIVE BACKFILLING SOIL		
UNIT OF EXPRESSION: km	REFERENCE DATE: *[dd.mm]*.yy	TYPE: Alpha-numeric
DEFINITION: Length of mains laid in trenches backfilled with very aggressive soil		
COMMENT: The exact definition of very aggressive soil shall be decided by the undertaking		

C44a – MAINS WITH MEDIUM AGGRESSIVE BACKFILLING SOIL		
UNIT OF EXPRESSION: km	REFERENCE DATE: *[dd.mm]*.yy	TYPE: Alpha-numeric
DEFINITION: Length of mains laid in trenches backfilled with medium aggressive soil		
COMMENT: The exact definition of medium aggressive soil shall be decided by the undertaking		

C44b – MAINS WITH LOW AGGRESSIVE BACKFILLING SOIL		
UNIT OF EXPRESSION: km	REFERENCE DATE: *[dd.mm]*.yy	TYPE: Alpha-numeric
DEFINITION: Length of mains laid in trenches backfilled with low aggressive soil		
COMMENT: The exact definition of low aggressive soil shall be decided by the undertaking		

C44c – MAINS WITH NOT AGGRESSIVE BACKFILLING SOIL		
UNIT OF EXPRESSION: km	REFERENCE DATE: *[dd.mm]*.yy	TYPE: Alpha-numeric
DEFINITION: Length of mains laid in trenches backfilled with not aggressive soil		
COMMENT: The exact definition of not aggressive soil shall be decided by the undertaking		

C45 - AVERAGE CLOSENESS TO TREES		
UNIT OF EXPRESSION: m	REFERENCE DATE: *[dd.mm]*.yy	TYPE: Real
DEFINITION: Average closeness between the pipe and the trees		
COMMENT:		

C46 - PIPE RIGID JOINTS		
UNIT OF EXPRESSION: km	REFERENCE DATE: *[dd.mm]*.yy	TYPE: Real
DEFINITION: Length of mains with rigid joints		
COMMENT:		

C47 - PIPE FLEXIBLE JOINTS		
UNIT OF EXPRESSION: km	REFERENCE DATE: *[dd.mm]*.yy	TYPE: Real
DEFINITION: Length of mains with flexible joints		
COMMENT:		

C29 - MAINS VALVES (IWA PI variable)		
UNIT OF EXPRESSION: No.	REFERENCE DATE: *[dd.mm]*.yy	TYPE: Integer
DEFINITION: Total number of all kinds of valves installed in transmission and distribution system mains.		
COMMENT: Valves installed in the service connections shall not be accounted for.		

C30 - ISOLATING VALVES (IWA PI variable)		
UNIT OF EXPRESSION: No.	REFERENCE DATE: *[dd.mm]*.yy	TYPE: Integer
DEFINITION: Total number of isolating valves of all types installed in transmission and distribution system mains.		
COMMENT: Valves installed in the service connections shall not be accounted for.		

C31 - HYDRANTS (IWA PI variable)		
UNIT OF EXPRESSION: No.	REFERENCE DATE: *[dd.mm]*.yy	TYPE: Integer
DEFINITION: Total number of hydrants of all types installed in transmission and distribution system.		
COMMENT:		

WATER STORAGE

C2 - TRANSMISSION & DISTRIBUTION STORAGE TANKS CAPACITY (IWA PI variable)		
UNIT OF EXPRESSION: m^3	REFERENCE DATE: *[dd.mm]*.yy	TYPE: Real
DEFINITION: Total volume of transmission and distribution storage tanks.		
COMMENT: The customer storage tanks must not be included. If appropriate, this variable can be split into capacity for fire flow protection and remaining capacity.		

PUMPING STATIONS

C4 - PUMPING STATIONS (IWA PI variable)		
UNIT OF EXPRESSION: No.	REFERENCE DATE: *[dd.mm]*.yy	TYPE: Integer
DEFINITION: Total number of pumping stations of the system (treatment plants and customer pumping systems excluded).		
COMMENT:		

PHYSICAL ASSETS DATA - SERVICE CONNECTIONS

C32 – NUMBER OF SERVICE CONNECTIONS (IWA PI variable)		
UNIT OF EXPRESSION: No.	REFERENCE DATE: *[dd.mm]*.yy	TYPE: Integer
DEFINITION: Total number of service connections.		
COMMENT:		

C32a **- NUMBER OF SENSITIVE SERVICE CONNECTIONS** (subset of the IWA PI variable C32)		
UNIT OF EXPRESSION: No.	REFERENCE DATE: *[dd.mm]*.yy	TYPE: Integer
DEFINITION: Total number of sensitive service connections, related with special consumer strongly dependent from the water supply (hospitals, schools, etc.) Sensitive service connections can be defined as connections where serious human consequences may occur due to supply interruptions (for example hospitals, dialysis patients, etc).		
COMMENT:		

C33 - AVERAGE SERVICE CONNECTION LENGTH TO MEASUREMENT POINT (IWA PI variable)		
UNIT OF EXPRESSION: m	REFERENCE DATE: *[dd.mm]*.yy	TYPE: Real
DEFINITION: Average length of pipe between the mains and the measurement point.		
COMMENT: Frequently water undertakings do not have detailed accurate information to assess the service connection length. In these cases, a qualitative assessment will be adopted.		

C48a - PLASTIC SERVICE CONNECTIONS (IWA CI)		
UNIT OF EXPRESSION: %	REFERENCE DATE: *[dd.mm]*.yy	TYPE: Real
DEFINITION: Percentage of the total number of plastic service connections regarding the total number of service connections		
COMMENT: Frequently water undertakings do not have detailed accurate information to assess the service connection length. In these cases, a qualitative assessment will be adopted.		

C48b - STEEL SERVICE CONNECTIONS (IWA CI)		
UNIT OF EXPRESSION: %	REFERENCE DATE: *[dd.mm]*.yy	TYPE: Real
DEFINITION: Percentage of the total number of steel service connections regarding the total number of service connections		
COMMENT: Frequently water undertakings do not have detailed accurate information to assess the service connection length. In these cases, a qualitative assessment will be adopted.		

C48c - LEAD SERVICE CONNECTIONS (IWA CI)		
UNIT OF EXPRESSION: %	REFERENCE DATE: *[dd.mm]*.yy	TYPE: Real
DEFINITION: Percentage of the total number of lead service connections regarding the total number of service connections		
COMMENT: Frequently water undertakings do not have detailed accurate information to assess the service connection length. In these cases, a qualitative assessment will be adopted.		

C48d - OTHER MATERIALS SERVICE CONNECTIONS (IWA CI)		
UNIT OF EXPRESSION: %	REFERENCE DATE: *[dd.mm]*.yy	TYPE: Real
DEFINITION: Percentage of the total number of other material service connections regarding the total number of service connections		
COMMENT: Frequently water undertakings do not have detailed accurate information to assess the service connection length. In these cases, a qualitative assessment will be adopted.		

WATER VOLUME DATA

A1 - ANNUAL ABSTRACTION CAPACITY (IWA PI variable)		
UNIT OF EXPRESSION: m^3/year	PERIOD: *[dd.mm]*.yy-1 – *[dd.mm]*.yy	TYPE: Real
DEFINITION: Maximum annual allowance of water abstraction for water supply, based on the availability of water resources.		
COMMENT: If the maximum annual abstraction capacity is not clearly established as an allowance, it shall be estimated as accurately as possible.		

A2 - IMPORTED WATER ALLOWANCE (IWA PI variable)		
UNIT OF EXPRESSION: m^3/year	PERIOD: *[dd.mm]*.yy-1 – *[dd.mm]*.yy	TYPE: Real
DEFINITION: Maximum allowance of raw and treated water importation.		
COMMENT: If the maximum annual allowance is not contracted with the supplier, it shall be estimated as accurately as possible.		

A4 - WATER ABSTRACTED (IWA PI variable)		
UNIT OF EXPRESSION: m^3/year	PERIOD: *[dd.mm]*.yy-1 – *[dd.mm]*.yy	TYPE: Real
DEFINITION: The annual volume of water obtained for input to water treatment plants (or directly to the transmission and distribution systems) that were abstracted from raw water sources.		
COMMENT:		

A5 - IMPORTED RAW WATER (IWA PI variable)		
UNIT OF EXPRESSION: m^3/year	PERIOD: *[dd.mm]*.yy-1 – *[dd.mm]*.yy	TYPE: Real
DEFINITION: Total annual volume of raw water transferred from other water supply systems.		
COMMENT:		

A6 - EXPORTED RAW WATER (IWA PI variable)		
UNIT OF EXPRESSION: m^3/year	PERIOD: *[dd.mm]*.yy-1 – *[dd.mm]*.yy	TYPE: Real
DEFINITION: Total annual volume of raw water transferred to other water supply systems.		
COMMENT:		

A7 - WATER PRODUCED (IWA PI variable)		
UNIT OF EXPRESSION: m^3/year	PERIOD: *[dd.mm]*.yy-1 – *[dd.mm]*.yy	TYPE: Real
DEFINITION: Total annual volume of water treated for input to water transmission lines or directly to the distribution system.		
COMMENT: Imported treated water should not be included.		

A8 - IMPORTED TREATED WATER IWA PI variable)		
UNIT OF EXPRESSION: m^3/year	PERIOD: *[dd.mm]*.yy-1 – *[dd.mm]*.yy	TYPE: Real
DEFINITION: Total annual volume of treated water imported from other water undertaking or system from the supply area.		
COMMENT:		

A9 - EXPORTED TREATED WATER (IWA PI variable)		
UNIT OF EXPRESSION: m^3/year	PERIOD: *[dd.mm]*.yy-1 – *[dd.mm]*.yy	TYPE: Real
DEFINITION: Total annual volume of treated water exported to other water undertaking or system from the supply area.		
COMMENT: These transfers can occur anywhere downstream of the treatment plants.		

A19 - AUTHORISED CONSUMPTION (IWA PI variable)		
UNIT OF EXPRESSION: m^3/year	PERIOD: *[dd.mm]*.yy-1 – *[dd.mm]*.yy	TYPE: Real
DEFINITION: Total annual volume of metered and/or non-metered water taken by registered customers, the water supplier and others who are implicitly or explicitly authorised to do so by the water supplier, for residential, commercial and industrial purposes. It includes water exported.		
COMMENT: Note that authorised consumption may include items such as fire fighting and training, flushing of mains and sewers, street cleaning, watering of municipal gardens, public fountains, frost protection, building water, etc. These may be billed or unbilled, metered or unmetered, according to local practice.		

A20 - WATER LOSSES (IWA PI variable)		
UNIT OF EXPRESSION: m^3/year	PERIOD: *[dd.mm]*.yy-1 – *[dd.mm]*.yy	TYPE: Real
DEFINITION: The difference between SYSTEM INPUT VOLUME and AUTHORISED CONSUMPTION.		
COMMENT: Water losses can be considered as a total volume for the whole system, or for partial systems such as raw water mains, transmission or distribution. In each case the components of the calculation would be adjusted accordingly. Water losses consist of real and apparent losses.		

A24 - REAL LOSSES (IWA PI variable)		
UNIT OF EXPRESSION: m^3/year	PERIOD: *[dd.mm]*.yy-1 – *[dd.mm]*.yy	TYPE: Real
DEFINITION: Total annual amount of physical water losses from the pressurised system, up to the point of customer metering.		
COMMENT: The annual volume lost through all types of leaks, bursts and overflows depends on frequencies, flow rates, and average duration of individual leaks.		

OPERATIONAL DATA

SERVICE PRESSURE

D31 - AVERAGE OPERATING PRESSURE (IWA PI variable and IWA CI)		
UNIT OF EXPRESSION: - kPa	PERIOD: *[dd.mm]*.yy-1 – *[dd.mm]*.yy	TYPE: Real
DEFINITION: Average operating pressure at the delivery point when system is pressurised.		
COMMENT: An exact measure of the average operating pressure would require the continuous monitoring of the pressure at every delivery point. In practice, simplifications are required. In flat regions, an estimate of this variable is easy to get. Also when calibrated extended period hydraulic models are available, nodal pressures weighted with the nodal demand can be adopted. In hilly areas where no better estimates can be obtained, simplified pressure contour maps can be drawn and an estimate for the average operating pressure in each contour band is established. The global value is than assessed as a weighted-average of the pressure at each contour band, using their respective equivalent-population as weighting factor.		

D31a - PRESSURE VARIATION (IWA CI)		
UNIT OF EXPRESSION: - kPa	PERIOD: *[dd.mm]*.yy-1 – *[dd.mm]*.yy	TYPE: Real
DEFINITION: Daily maximum difference between pressure and dynamic pressure in every delivery point, not including hydraulic transient situations		
COMMENT:		

SERVICE CONTINUITY

D29 - TIME SYSTEM IS PRESSURISED (IWA PI variable)		
UNIT OF EXPRESSION: hour/year	PERIOD: *[dd.mm]*.yy-1 – *[dd.mm]*.yy	TYPE: Real
DEFINITION: Amount of time of the year the system is pressurized.		
COMMENT: Interruptions due to unplanned system failures, or to on-going repair or rehabilitation works shall not be accounted for by this variable. In most intermittent supply systems, supply interruptions are not simultaneous all over the network. When there are subsystems supplied in different periods, the indicator has to be assessed individually for each subsystem and the result is a weighted average using the number of service connections of each subsystem as weighting factor.		

D30 - DELIVERY POINTS WITH ADEQUATE PRESSURE (IWA PI variable)		
UNIT OF EXPRESSION: No./year	PERIOD: *[dd.mm]*.yy-1 – *[dd.mm]*.yy	TYPE: Integer
DEFINITION: Number of delivery points that receive and are likely to receive pressure equal to or above the guaranteed or declared target level at the peak demand hour (but not when demand is abnormal).		
COMMENT:		

D32 - WATER INTERRUPTIONS (IWA PI variable)		
UNIT OF EXPRESSION: No..h/year	PERIOD: *[dd.mm]*.yy-1 – *[dd.mm]*.yy	TYPE: Integer
DEFINITION: Σ (Population subject to a water interruption x duration of the interruption in hours).		
COMMENT:		

D33 - SERVICE INTERRUPTIONS (IWA PI variable)		
UNIT OF EXPRESSION: No./year	PERIOD: *[dd.mm]*.yy-1 – *[dd.mm]*.yy	TYPE: Integer
DEFINITION: Total number of service interruptions.		
COMMENT: Interruptions inherent to a systematic intermittent supply must not be accounted in this variable.		

D33a - CRITICAL INTERRUPTIONS (subset of the IWA PI variable D33)		
UNIT OF EXPRESSION: No./year	PERIOD: *[dd.mm]*.yy-1 – *[dd.mm]*.yy	TYPE: Integer
DEFINITION: Total number of supply interruptions to sensitive service connections during the year.		
COMMENT: Sensitive service connections can be defined as connections where serious human consequences may occur due to supply interruptions (for example hospitals, dialysis patients, etc). Interruptions inherent to a systematic intermittent supply must not be accounted in this variable.		

D34 - WATER USE RESTRICTIONS IWA PI variable)		
UNIT OF EXPRESSION: No. .h / year	PERIOD: *[dd.mm]*.yy-1 – *[dd.mm]*.yy	TYPE: Real
DEFINITION: Σ (Population affected by restrictions to water service x duration of the restrictions to water service in hours).		
COMMENT:		

D35 - DAYS WITH RESTRICTIONS TO WATER SERVICE (IWA PI variable)		
UNIT OF EXPRESSION: days/year	PERIOD: *[dd.mm]*.yy-1 – *[dd.mm]*.yy	TYPE: Integer
DEFINITION: Total number of days with restrictions to water service during the year.		
COMMENT:		

WATER QUALITY MONITORING

D41 - WATER QUALITY TESTS PERFORMED (IWA PI variable)		
UNIT OF EXPRESSION: No./year	PERIOD: *[dd.mm]*.yy-1 – *[dd.mm]*.yy	TYPE: Integer
DEFINITION: Number of treated water tests performed during the year.		
COMMENT:		

D42 - AESTHETIC TESTS PERFORMED (IWA PI variable)		
UNIT OF EXPRESSION: No./year	PERIOD: *[dd.mm]*.yy-1 – *[dd.mm]*.yy	TYPE: Integer
DEFINITION: Number of aesthetic tests of treated water performed during the year.		
COMMENT:		

D42a - TASTE TESTS PERFORMED (subset of the IWA PI variable D42)		
UNIT OF EXPRESSION: No./year	PERIOD: *[dd.mm]*.yy-1 – *[dd.mm]*.yy	TYPE: Integer
DEFINITION: Number of taste tests of treated water performed during the year.		
COMMENT:		

D42b - COLOUR TESTS PERFORMED (subset of the IWA PI variable D42)		
UNIT OF EXPRESSION: No./year	PERIOD: *[dd.mm]*.yy-1 – *[dd.mm]*.yy	TYPE: Integer
DEFINITION: Number of colour tests of treated water performed during the year.		
COMMENT:		

D43 - MICROBIOLOGICAL TESTS PERFORMED (IWA PI variable)		
UNIT OF EXPRESSION: No./year	PERIOD: *[dd.mm]*.yy-1 – *[dd.mm]*.yy	TYPE: Integer
DEFINITION: Number of microbiological tests of treated water performed during the year.		
COMMENT:		

D44 - PHYSICAL-CHEMICAL TESTS PERFORMED (IWA PI variable)		
UNIT OF EXPRESSION: No./year	PERIOD: *[dd.mm]*.yy-1 – *[dd.mm]*.yy	TYPE: Integer
DEFINITION: Number of physical-chemical tests of treated water performed during the year.		
COMMENT:		

D51 - COMPLIANCE OF AESTHETIC TESTS (IWA PI variable)		
UNIT OF EXPRESSION: No./year	PERIOD: *[dd.mm]*.yy-1 – *[dd.mm]*.yy	TYPE: Integer
DEFINITION: Number of aesthetic tests of treated water performed during the year and complying with the applicable standards or legislation.		
COMMENT:		

D51a - COMPLIANCE OF WATER TASTE TESTS (subset of the IWA PI variable D51)		
UNIT OF EXPRESSION: No./year	PERIOD: *[dd.mm]*.yy-1 – *[dd.mm]*.yy	TYPE: Integer
DEFINITION: Number of water taste tests of treated water performed during the year and complying with the applicable standards or legislation.		
COMMENT:		

D51b - COMPLIANCE OF WATER COLOUR TESTS (subset of the IWA PI variable D51)		
UNIT OF EXPRESSION: No./year	PERIOD: *[dd.mm]*.yy-1 – *[dd.mm]*.yy	TYPE: Integer
DEFINITION: Number of water colour tests of treated water performed during the year and complying with the applicable standards or legislation.		
COMMENT:		

D52 - COMPLIANCE OF MICROBIOLOGICAL TESTS (IWA PI variable)		
UNIT OF EXPRESSION: No./year	PERIOD: *[dd.mm]*.yy-1 – *[dd.mm]*.yy	TYPE: Integer
DEFINITION: Number of microbiological tests of treated water performed during a year and complying with the applicable standards or legislation.		
COMMENT:		

D53 - COMPLIANCE OF PHYSICAL-CHEMICAL TESTS (IWA PI variable)		
UNIT OF EXPRESSION: No./year	PERIOD: *[dd.mm]*.yy-1 – *[dd.mm]*.yy	TYPE: Integer
DEFINITION: Number of physical-chemical tests of treated water performed during a year and complying with the applicable standards or legislation.		
COMMENT:		

INSPECTION AND MAINTENANCE

D7 - NETWORK INSPECTION (IWA PI variable)		
UNIT OF EXPRESSION: km/year	PERIOD: *[dd.mm]*.yy-1 – *[dd.mm]*.yy	TYPE: Real
DEFINITION: Length of transmission and distribution mains where at least valves and other fittings inspected during the year.		
COMMENT:		

D9 - LEAKS REPAIRED DUE TO ACTIVE LEAKAGE CONTROL (IWA PI variable)		
UNIT OF EXPRESSION: No./year	PERIOD: *[dd.mm]*.yy-1 – *[dd.mm]*.yy	TYPE: Real
DEFINITION: Number of mains leaks detected and repaired due to active leakage control.		
COMMENT:		

PREVENTIVE MAINTENANCE

D18 - MAINS REHABILITATION (IWA PI variable)		
UNIT OF EXPRESSION: km/year	PERIOD: *[dd.mm]*.yy-1 – *[dd.mm]*.yy	TYPE: Real
DEFINITION: Length of transmission and distribution mains rehabilitated during the year.		
COMMENT: This variable includes not only D19 and D20 but also the length of mains rehabilitated with other techniques.		

D19 - MAINS RELINING (IWA PI variable)		
UNIT OF EXPRESSION: km/year	PERIOD: *[dd.mm]*.yy-1 – *[dd.mm]*.yy	TYPE: Real
DEFINITION: Length of mains relined during the year by epoxy resin or cement mortar.		
COMMENT:		

D20 - REPLACED OR RENEWED MAINS (IWA PI variable)		
UNIT OF EXPRESSION: km/year	PERIOD: *[dd.mm]*.yy-1 – *[dd.mm]*.yy	TYPE: Real
DEFINITION: Mains length replaced during the year or renewed by trenchless techniques.		
COMMENT:		

D21 - REPLACED OR RENEWED VALVES (IWA PI variable)		
UNIT OF EXPRESSION: No./year	PERIOD: *[dd.mm]*.yy-1 – *[dd.mm]*.yy	TYPE: Integer
DEFINITION: Number of mains valves replaced or renewed during the year.		
COMMENT: Valves installed in service connections shall not be included.		

D22 - SERVICE CONNECTION REHABILITATION (IWA PI variable)		
UNIT OF EXPRESSION: No./year	PERIOD: *[dd.mm]*.yy-1 – *[dd.mm]*.yy	TYPE: Integer
DEFINITION: Number of service connections replaced or renewed during the year.		
COMMENT:		

FAILURES

D25 - MAINS FAILURES (IWA PI variable)		
UNIT OF EXPRESSION: No./year	PERIOD: *[dd.mm]*.yy-1 – *[dd.mm]*.yy	TYPE: Integer
DEFINITION: Number of mains failures during the year, including failures of valves and fittings.		
COMMENT:		

D25a - PIPE FAILURES (subset of the IWA PI variable D25)		
UNIT OF EXPRESSION: No./year	PERIOD: *[dd.mm]*.yy-1 – *[dd.mm]*.yy	TYPE: Integer
DEFINITION: Number of pipe failures during the year, excluding failures of valves joints and links to service connections.		
COMMENT:		

D25b - JOINT FAILURES (subset of the IWA PI variable D25)		
UNIT OF EXPRESSION: No./year	PERIOD: *[dd.mm]*.yy-1 – *[dd.mm]*.yy	TYPE: Integer
DEFINITION: Number of joint failures during the year.		
COMMENT:		

D25c - VALVE FAILURES (subset of the IWA PI variable D25)		
UNIT OF EXPRESSION: No./year	PERIOD: *[dd.mm]*.yy-1 – *[dd.mm]*.yy	TYPE: Integer
DEFINITION: Number of valve failures during the year.		
COMMENT:		

D25d - SERVICE CONNECTION INSERTION POINT FAILURES (subset of the IWA PI variable D25)		
UNIT OF EXPRESSION: No./year	PERIOD: *[dd.mm]*.yy-1 – *[dd.mm]*.yy	TYPE: Integer
DEFINITION: Number of service connections failures that occur in the insertion point of the service connection during the year.		
COMMENT: If service connections failures are to be used for regulating objectives, the use of a complementary indicator, similar to *Op26* but excluding failures by third parties is advisable, as they are not a direct fault of the water undertaking. Number should exclude repairs under active leakage control. Number should exclude repairs under active leakage control.		

D25e - MAINS FAILURES CAUSED BY THIRD PARTIES (subset of the IWA PI variable D25)		
UNIT OF EXPRESSION: No./year	PERIOD: *[dd.mm]*.yy-1 – *[dd.mm]*.yy	TYPE: Integer
DEFINITION: Number of mains failures during the year caused by third parties, e.g. gas utility, TV cable utility, etc.		
COMMENT:		

D25f - CRITICAL MAINS FAILURES (subset of the IWA PI variable D25)		
UNIT OF EXPRESSION: No./year	PERIOD: *[dd.mm]*.yy-1 – *[dd.mm]*.yy	TYPE: Integer
DEFINITION: Number of critical mains failures during the year, including failures of valves and fittings.		
COMMENT: Critical main failures can be defined where serious human consequences may occur due to supply interruptions (for example hospitals, dialysis patients, etc). Sensitive areas can be defined as urban zones where serious human consequences (for example hospitals, dialysis patients, etc), severe damages (landslide, flooding) or severe disturbances (traffic interruptions) may occur due to water main bursts.		

D26 - SERVICE CONNECTION FAILURES (IWA PI variable)		
UNIT OF EXPRESSION: No./year	PERIOD: *[dd.mm]*.yy-1 – *[dd.mm]*.yy	TYPE: Integer
DEFINITION: Number of service connections failures during the year.		
COMMENT: If service connections failures are to be used for regulating objectives, the use of a complementary indicator, similar to *Op26* but excluding failures by third parties is advisable, as they are not a direct fault of the water undertaking. Number should exclude repairs under active leakage control. Number should exclude repairs under active leakage control.		

D27 - HYDRANT FAILURES (IWA PI variable)		
UNIT OF EXPRESSION: No./year	PERIOD: *[dd.mm]*.yy-1 – *[dd.mm]*.yy	TYPE: Integer
DEFINITION: Number of hydrant failures during the year.		
COMMENT: If hydrant failures are to be used for regulating objectives, the use of a complementary indicator, similar to *Op28* but excluding failures by third parties is advisable, as they are not a direct fault of the water undertaking. Number should exclude repairs under active leakage control.		

D28 - POWER FAILURES (IWA PI variable)		
UNIT OF EXPRESSION: hour/year	PERIOD: *[dd.mm]*.yy-1 – *[dd.mm]*.yy	TYPE: Real
DEFINITION: $\Sigma_{(\text{for all pumping stations})}$ (number of hours during the year each pumping station is out of service or is reliant on standby power generation due to power supply interruptions).		
COMMENT:		

REHABILITATION

D36 - MAINS REHABILITATED		
UNIT OF EXPRESSION: km/year	PERIOD: *[dd.mm]*.yy-1 – *[dd.mm]*.yy	TYPE: Real
DEFINITION: Length of mains rehabilitated per year		
COMMENT:		

D36a - MAINS RELINED		
UNIT OF EXPRESSION: km/year	PERIOD: *[dd.mm]*.yy-1 – *[dd.mm]*.yy	TYPE: Real
DEFINITION: Length of mains relined per year		
COMMENT:		

D36b - MAINS REPLACED OR RENEWED		
UNIT OF EXPRESSION: km/year	PERIOD: *[dd.mm]*.yy-1 – *[dd.mm]*.yy	TYPE: Real
DEFINITION: Length of mains replaced or renewed per year		
COMMENT:		

D36c - VALVES REPLACED OR RENEWED		
UNIT OF EXPRESSION: No./year	PERIOD: *[dd.mm]*.yy-1 – *[dd.mm]*.yy	TYPE: Real
DEFINITION: Number of valves replaced or renewed per year		
COMMENT:		

D37 - MAINS REHABILITATION AVERAGE DATE		
UNIT OF EXPRESSION: year	PERIOD: *[dd.mm]*.yy-1 – *[dd.mm]*.yy	TYPE: Real
DEFINITION: Average date of the rehabilitated mains		
COMMENT:		

D37a - MAINS RELINED AVERAGE DATE		
UNIT OF EXPRESSION: year	PERIOD: *[dd.mm]*.yy-1 – *[dd.mm]*.yy	TYPE: Real
DEFINITION: Average date of the relined mains		
COMMENT:		

D37b - MAINS REPLACED OR RENEWED AVERAGE DATE		
UNIT OF EXPRESSION: year	PERIOD: *[dd.mm]*.yy-1 – *[dd.mm]*.yy	TYPE: Real
DEFINITION: Average date of the replaced or renewed mains		
COMMENT:		

D37c - VALVES REPLACED OR RENEWED AVERAGE DATE		
UNIT OF EXPRESSION: year	PERIOD: *[dd.mm]*.yy-1 – *[dd.mm]*.yy	TYPE: Real
DEFINITION: Average date of the replaced or renewed valves		
COMMENT:		

D38 - SERVICE CONNECTIONS REHABILITATED		
UNIT OF EXPRESSION: No./year	PERIOD: *[dd.mm]*.yy-1 – *[dd.mm]*.yy	TYPE: Real
DEFINITION: Number of service connections rehabilitated per year		
COMMENT:		

D39 - SERVICE CONNECTIONS REHABILITATION AVERAGE DATE		
UNIT OF EXPRESSION: year	PERIOD: *[dd.mm]*.yy-1 – *[dd.mm]*.yy	TYPE: Real
DEFINITION: Average date of the rehabilitated service connections		
COMMENT:		

QUALITY OF SERVICE DATA

F1 - POPULATION SUPPLIED (IWA PI variable)		
UNIT OF EXPRESSION: No.	REFERENCE DATE: *[dd.mm]*.yy	TYPE: Integer
DEFINITION: Resident population served by the water undertaking.		
COMMENT:		

F11 - SERVICE CUSTOMER COMPLAINTS (IWA PI variable)		
UNIT OF EXPRESSION: No./year	PERIOD: *[dd.mm]*.yy-1 – *[dd.mm]*.yy	TYPE: Integer
DEFINITION: Number of direct, telephone, and written complaints of quality of service during the year.		
COMMENT:		

F12 - PRESSURE CUSTOMER COMPLAINTS (IWA PI variable)		
UNIT OF EXPRESSION: No./year	PERIOD: *[dd.mm]*.yy-1 – *[dd.mm]*.yy	TYPE: Integer
DEFINITION: Number of pressure complaints during the year.		
COMMENT:		

F13 - CONTINUITY CUSTOMER COMPLAINTS (IWA PI variable)		
UNIT OF EXPRESSION: No./year	PERIOD: *[dd.mm]*.yy-1 – *[dd.mm]*.yy	TYPE: Integer
DEFINITION: Number of continuity complaints during the year.		
COMMENT: This variable refers to medium- or long-term supply constraints. Complaints on supply interruptions or water use restrictions due to insufficient water quantity or quality at source, insufficient system capacity or on-going works shall be accounted for by this variable. Any other complaints on interruptions shall be included in F.		

F14 - WATER QUALITY CUSTOMER COMPLAINTS (IWA PI variable)		
UNIT OF EXPRESSION: No./year	PERIOD: *[dd.mm]*.yy-1 – *[dd.mm]*.yy	TYPE: Integer
DEFINITION: Number of water quality complaints during the year.		
COMMENT:		

F14a - WATER TASTE CUSTOMER COMPLAINTS (subset of the IWA PI variable F14)		
UNIT OF EXPRESSION: No./year	PERIOD: *[dd.mm]*.yy-1 – *[dd.mm]*.yy	TYPE: Integer
DEFINITION: Number of water taste customer complaints during the year.		
COMMENT:		

F14b - WATER COLOUR CUSTOMER COMPLAINTS (subset of the IWA PI variable F14)		
UNIT OF EXPRESSION: No./year	PERIOD: *[dd.mm]*.yy-1 – *[dd.mm]*.yy	TYPE: Integer
DEFINITION: Number of water colour customer complaints during the year.		
COMMENT:		

F15 - CUSTOMER COMPLAINTS ON INTERRUPTIONS (IWA PI variable F15)		
UNIT OF EXPRESSION: No./year	PERIOD: *[dd.mm]*.yy-1 – *[dd.mm]*.yy	TYPE: Integer
DEFINITION: Number of customer complaints due to supply interruptions during the year.		
COMMENT: This variable refers to short-term supply interruptions, due to accidental system failures or repair works. Supply interruptions due to intermittent supply, insufficient quantity of adequate water at sources, or asset insufficient capacity shall be accounted for in F and not in F.		

F15a - CUSTOMER COMPLAINTS ON CRITICAL INTERRUPTIONS (subset of the IWA PI variable F15)		
UNIT OF EXPRESSION: No./year	PERIOD: *[dd.mm]*.yy-1 – *[dd.mm]*.yy	TYPE: Integer
DEFINITION: Number of complaints due to supply interruptions in critical areas during the year.		
COMMENT: This variable refers to short-term supply interruptions, due to accidental system failures or repair works. Supply interruptions due to intermittent supply, insufficient quantity of adequate water at sources, or asset insufficient capacity shall be accounted for in F and not in F. Sensitive service connections can be defined as connections where serious human consequences may occur due to supply interruptions (for example hospitals, dialysis patients, etc).		

FINANCIAL DATA

G1 - ANNUAL COSTS (IWA PI variable)		
UNIT OF EXPRESSION: €/year	PERIOD: *[dd.mm]*.yy-1 – *[dd.mm]*.yy	TYPE: Real
DEFINITION: Total annual costs, including capital and running costs.		
PROCESSING RULE: G2+G3		
COMMENT: Exchange rate of local currencies shall be referred to the end of the year.		

G2 - ANNUAL RUNNING COSTS (IWA PI variable)		
UNIT OF EXPRESSION: €/year	PERIOD: *[dd.mm]*.yy-1 – *[dd.mm]*.yy	TYPE: Real
DEFINITION: Total annual operations and maintenance costs + internal manpower costs - capitalised costs of self-constructed assets.		
COMMENT: This definition has, on aggregate level, to be equivalent to the sum of the NET desegregated values allocated at the numerator of the indicators figuring the composition of annual running costs per type of cost: Fi4, Fi5, Fi6, Fi7 and Fi8. Exchange rate of local currencies shall be referred to the end of the year.		

G3 - ANNUAL CAPITAL COSTS (IWA PI variable)		
UNIT OF EXPRESSION: €/year	PERIOD: *[dd.mm]*.yy-1 – *[dd.mm]*.yy	TYPE: Real
DEFINITION: Total annual net interest and depreciation (based on book values).		
COMMENT: The net value of interest has to be considered, being the interest income a reduction in capital costs and not a revenue. Exchange rate of local currencies shall be referred to the end of the year.		

G19 - ANNUAL DEPRECIATION COSTS (IWA PI variable)		
UNIT OF EXPRESSION: €/year	PERIOD: *[dd.mm]*.yy-1 – *[dd.mm]*.yy	TYPE: Real
DEFINITION: Annual depreciation costs (on book values).		
COMMENT: Exchange rate of local currencies shall be referred to the end of the year.		

G26 - ANNUAL INVESTMENT IN TANGIBLE ASSETS (IWA PI variable)		
UNIT OF EXPRESSION: €/year	PERIOD: *[dd.mm]*.(yy-1) - *[dd.mm]*.yy	TYPE: Real
DEFINITION: Total cost of the investments in tangible (expenditures for plant and equipment), including capitalised cost of self-constructed tangible assets (apportionment of G29 as for related to tangible assets).		
COMMENT: Tangible assets include investment for supporting buildings, vehicles, etc. Exchange rate of local currencies shall be referred to the end of the year.		

G27 - ANNUAL INVESTMENTS FOR NEW ASSETS (IWA PI variable)		
UNIT OF EXPRESSION: €/year	PERIOD: *[dd.mm]*.yy-1 - *[dd.mm]*.yy	TYPE: Real
DEFINITION: Total cost of the investments in tangible assets that constitute a new development for the service, including capitalised cost of self-constructed new assets (apportionment of G29 as for related to tangible new assets).		
COMMENT: Exchange rate of local currencies shall be referred to the end of the year.		

G28 - ANNUAL INVESTMENTS FOR ASSETS REPLACEMENT (IWA PI variable)		
UNIT OF EXPRESSION: €/year	PERIOD: *[dd.mm]*.yy-1 - *[dd.mm]*.yy	TYPE: Real
DEFINITION: Total cost of the investments related to the existing assets (i.e. maintaining the existing infrastructure at the same level), including capitalised cost of self-constructed replaced assets (apportionment of G29 as for related to the replacement of tangible assets).		
COMMENT: Exchange rate of local currencies shall be referred to the end of the year.		

G21 - ANNUAL REVENUE (IWA PI variable)		
UNIT OF EXPRESSION: €/year	PERIOD: *[dd.mm]*.yy-1 – *[dd.mm]*.yy	TYPE: Real
DEFINITION: Total operating revenues – capitalised costs of self constructed assets		
PROCESSING RULE:	Input data	
COMMENT: Capitalised costs of self-constructed assets have more correctly to be intended not as revenue but as an economic correction of each type of cost to negative apportion. These capitalisations have consequently to be negative allocated even for the calculation of annual revenues. Exchange rate of local currencies shall be referred to the end of the year.		

G30 - WATER SALES REVENUE FOR DIRECT CONSUMPTION (IWA PI variable)		
UNIT OF EXPRESSION: €/year	PERIOD: *[dd.mm]*.yy-1 – *[dd.mm]*.yy	TYPE: Real
DEFINITION: Water sales revenue for direct consumption.		
COMMENT: Exchange rate of local currencies shall be referred to the end of the year.		

G31 - WATER SALES REVENUE FOR EXPORTED WATER (IWA PI variable)		
UNIT OF EXPRESSION: €/year	PERIOD: *[dd.mm]*.yy-1 – *[dd.mm]*.yy	TYPE: Real
DEFINITION: Water sales revenue for exported water.		
COMMENT: Exchange rate of local currencies shall be referred to the end of the year.		

G51 - PIPE REHABILITATION COST		
UNIT OF EXPRESSION: €/m	PERIOD: *[dd.mm]*.yy-1 – *[dd.mm]*.yy	TYPE: Real
DEFINITION: Average unit costs for pipe rehabilitation		
COMMENT: Exchange rate of local currencies shall be referred to the end of the year.		

G52 - PIPE REPAIR COST		
UNIT OF EXPRESSION: €/m	PERIOD: *[dd.mm]*.yy-1 – *[dd.mm]*.yy	TYPE: Real
DEFINITION: Average unit costs for pipe repairs		
COMMENT: Exchange rate of local currencies shall be referred to the end of the year.		

G53 - PIPE MAINTENANCE COST		
UNIT OF EXPRESSION: €/m	PERIOD: *[dd.mm]*.yy-1 – *[dd.mm]*.yy	TYPE: Real
DEFINITION: Average unit costs for pipe maintenance		
COMMENT: Exchange rate of local currencies shall be referred to the end of the year.		

G54 - PIPE INSPECTION COST		
UNIT OF EXPRESSION: €/m	PERIOD: *[dd.mm]*.yy-1 – *[dd.mm]*.yy	TYPE: Real
DEFINITION: Average unit costs for pipe inspection		
COMMENT: Exchange rate of local currencies shall be referred to the end of the year.		

G55 - REDUCED COST FROM PIPE REPAIR		
UNIT OF EXPRESSION: €/year	PERIOD: *[dd.mm]*.yy-1 – *[dd.mm]*.yy	TYPE: Real
DEFINITION: Average operation cost reduction due to pipes repair		
COMMENT: The benefits are cumulative values up to the year t depending on the kind and intensity of rehab measures taken Exchange rate of local currencies shall be referred to the end of the year.		

G56 - REDUCED COST FROM LEAKAGE REPAIR		
UNIT OF EXPRESSION: €/year	PERIOD: *[dd.mm]*.yy-1 – *[dd.mm]*.yy	TYPE: Real
DEFINITION: Average operation cost reduction due to leakage control		
COMMENT: The benefits are cumulative values up to the year t depending on the kind and intensity of rehab measures taken Exchange rate of local currencies shall be referred to the end of the year.		

G57 - DOMESTIC WATER CONSUMPTION TARIFF (IWA CI)		
UNIT OF EXPRESSION: €/m^3	PERIOD: *[dd.mm]*.yy-1 – *[dd.mm]*.yy	TYPE: Real
DEFINITION: Average domestic water consumption tariff paid by the consumer to the water utility		
COMMENT: Exchange rate of local currencies shall be referred to the end of the year.		

G58 - INDUSTRIAL WATER CONSUMPTION TARIFF		
UNIT OF EXPRESSION: €/m^3	PERIOD: *[dd.mm]*.yy-1 – *[dd.mm]*.yy	TYPE: Real
DEFINITION: Average industrial water consumption tariff paid by the consumer to the water utility		
COMMENT: Exchange rate of local currencies shall be referred to the end of the year.		

G59 - PUBLIC WATER CONSUMPTION TARIFF		
UNIT OF EXPRESSION: €/m^3	PERIOD: *[dd.mm]*.yy-1 – *[dd.mm]*.yy	TYPE: Real
DEFINITION: Average public water consumption tariff		
COMMENT: Exchange rate of local currencies shall be referred to the end of the year.		

Appendix 4
CARE-W rehab external information

REHAB EXTERNAL INFORMATION

(These statistics relate to the area of service)

DATA	*(unit)*	CONCEPT
ENVIRONMENT		
Annual rainfall		(average for the past 30 years)
• *average*	*(mm/year)*	Annual average rainfall (average for the past 30 years)
• *maximum*	*(mm/year)*	Annual maximum rainfall assessed as the annual maxima of the last 30 years
• *minimum*	*(mm/year)*	Annual minimum rainfall assessed as the annual minima of the last 30 years
Air temperature		(average for the past 30 years)
• *daily average*	*(°C)*	Average daily air temperature of the year (averages for the past 30 years)
• *daily maximum*	*(°C)*	Average air temperature for the hottest day of the year (averages for the past 30 years)
• *daily minimum*	*(°C)*	Average air temperature for the coldest day of the year (averages for the past 30 years)
Topography		
• *source average elevation*	*(m)*	Weighted average elevation above sea level of water sources, including import delivery points, using source production as weighing factors
• *maximum delivery elevation*	*(m)*	Maximum elevation above sea level at the water delivery points of water distribution area
• *minimum delivery elevation*	*(m)*	Minimum elevation above sea level at the water delivery points of water distribution area
MAINS AGRESSIVE FACTORS		
Physical and chemical soil and groundwater characteristics		
Hydrogenionic concentration range	*(pH-pH)*	Maximum and minimum values of the soil and groundwater hydrogenionic concentration in the area around the network
Aggressive carbon dioxide concentration range (CO_2)	*(mg/l-mg/l)*	Maximum and minimum values of the soil and groundwater aggressive carbon dioxide concentration in the area around the network
Sulphate concentration range (SO_4^{2-})	*(mg/l-mg/l)*	Maximum and minimum values of the soil and groundwater sulphate concentration in the area around the network
Chloride concentration range (Cl^-)	*(mg/l-mg/l)*	Maximum and minimum values of the soil and groundwater chloride concentration in the area around the network
Resistivity concentration range	*(Ωm -Ωm)*	Maximum and minimum values of the soil and groundwater resistivity in the area around the network
Organic compounds	*(yes/no)*	Existence of organic compounds in the area around the network
Stray currents	*(yes/no)*	Existence of stray currents in the area around the network
Geotechnical conditions		
High seat stability	*(yes/no)*	Existence of high seat stability conditions of the soil below the pipes
Medium seat stability	*(yes/no)*	Existence of medium seat stability conditions of the soil below the pipes
Low seat stability	*(yes/no)*	Existence of low seat stability conditions of the soil below the pipes
Traffic class		
Heavy traffic	*(code HT)*	Heavy traffic conditions in the pavement above the pipes
Normal traffic	*(code NT)*	Normal traffic conditions in the pavement above the pipes
Light traffic	*(code LT)*	Light traffic conditions in the pavement above the pipes
Interference with other infrastructures		
High risk to be affected by other infrastructures works	*(yes/no)*	High risk of the pipes to be affected by other infrastructure works
Medium risk to be affected by other infrastructures works	*(yes/no)*	Medium risk of the pipes to be affected by other infrastructure works

DATA (unit)	CONCEPT
Low risk to be affected by other infrastructures works *(yes/no)*	Low risk of the pipes to be affected by other infrastructure works
Seismic conditions	
Range of maximum soil movement due to liquefaction *(mm)*	Forecasted maximum soil movement due to soil liquefaction in seismic conditions
Range of maximum angular deflection in joints *(°)*	Forecasted maximum angular deflection in pipe joints in seismic conditions
Range of maximum axial displacement in joints *(mm)*	Forecasted maximum axial displacement in pipe joints in seismic conditions
ECONOMICS	
Inflation rate *(%/year)*	Official annual inflation rate at the end of the year in the country

Appendix 5

Detailed results of the CARE-W PI comparison among end-users

1.1 Operational indicators

1.1.1 Rehabilitation

Op15 - MAINS REHABILITATION (%/year)

Concept: Length of transmission and distribution mains rehabilitated during the year / total mains length x 100

Processing rule: Op15 = D18/C6 x 100

Variables: D18 – Mains rehabilitation (km/year)
C6 – Mains length (km)

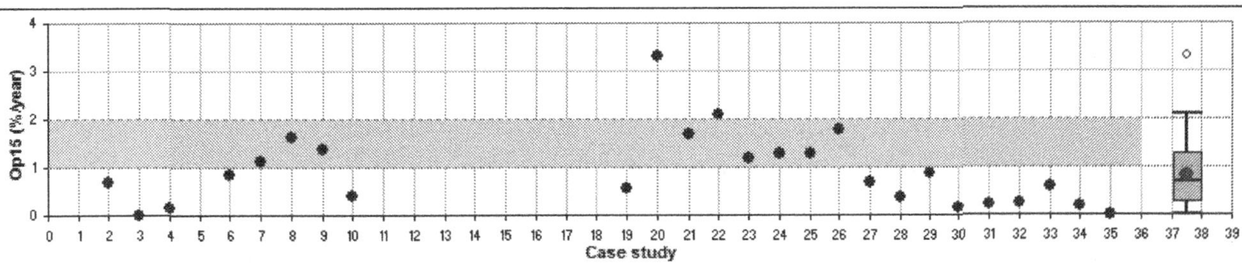

Comments: The mean value of the results collected is 0.91% of mains rehabilitated per year and 50% of the case studies are situated in the range from 0.25 to 1.3% of mains rehabilitated per year. As it might be expected, a significant number of case studies present relatively low rehabilitation rates, corresponding in average to a mains lifetime of 110 years.

Difficulties: A few end-users had difficulties with D18 (Mains rehabilitation). This difficulty seems to be due to the lack of record, not to the complexity of the variable itself, and nobody recommended abandoning this PI due to assessment difficulties.

Relevance: This PI is considered essential or important by all end-users.

PI guidance range: At this stage, a tentative recommended guidance range seems to be 1.0-2.0% of mains rehabilitated per year, representing a global network renovation within a period of 50 to 100 years.

Op16 - MAINS RENOVATION (%/year)

Concept: Length of mains renovated during the year / total mains length x 100

Processing rule: Op16 = D19/C6 x 100

Variables: D19 – Mains renovation (km/year)
C6 – Mains length (km)

Parent PI: Op15

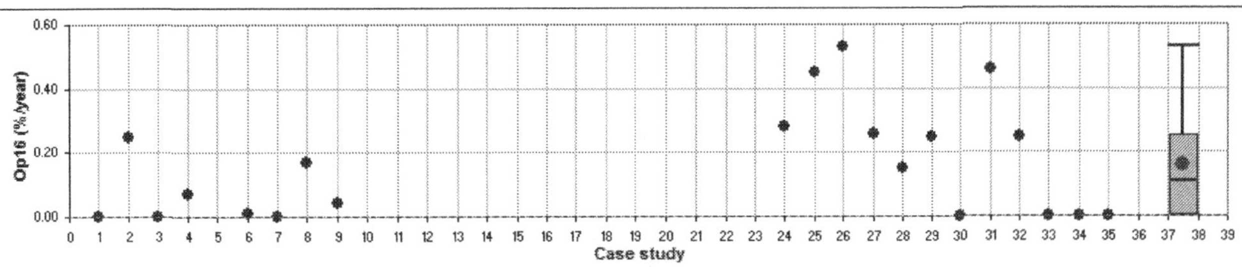

Comments: The mean value of the results collected is 0.15% of mains renovated per year and 50% are situated in the range from 0.0 to 0.26% of mains renovated per year. As it might be expected, a significant number of case studies present relatively low mains renovation values.

Difficulties: Several end-users had difficulties with D19 (Mains renovation). This difficulty seems to be due to the lack of record, not to the complexity of the variable itself, and nobody recommended abandoning this PI.

Relevance: This PI is considered essential or important by almost all end-users.

PI guidance range: It does not seem to be advisable to establish a guidance range for this indicator because the rehabilitation technology to be adopted is case depending.

Op17 - **MAINS REPLACEMENT** (%/year)

Concept: Length of mains replaced during the year / total mains length x 100

Processing rule: Op17 = D20/C6 x 100

Variables: D20 – Mains replacement (km/year)
C6 – Mains length (km)

Parent PI: Op15

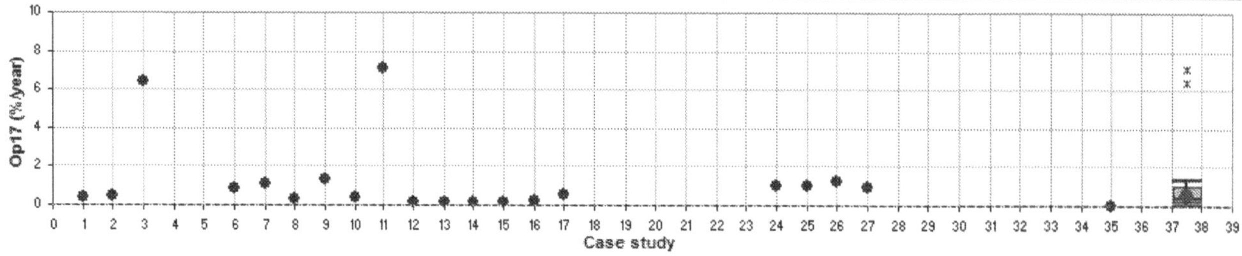

Comments: The mean value of the results collected is 1.2% of replaced or renewed mains per year and 50% are situated in the range from 0.23 to 1.0%. According to these results, there are much more mains replacement than renovated.

Difficulties: Several end-users had difficulties with D20 (Replaced or renewed mains). This difficulty seems to be due to the lack of record, not to the complexity of the variable itself, and nobody recommended abandoning this PI.

Relevance: This PI is considered essential or important almost by all end-users.

PI guidance range: It does not seem to be advisable to establish a guidance range for this indicator because the rehabilitation technology to be adopted is case depending.

Op18 - **VALVE REPLACEMENT** (%/year)

Concept: Number of mains valves replaced during the year / total number of mains valves x 100

Processing rule: Op18 = D21/C29 x 100

Variables: D21 – Replaced valves (No./year)
C29 – Mains valves (No.)

Parent PI: Op15

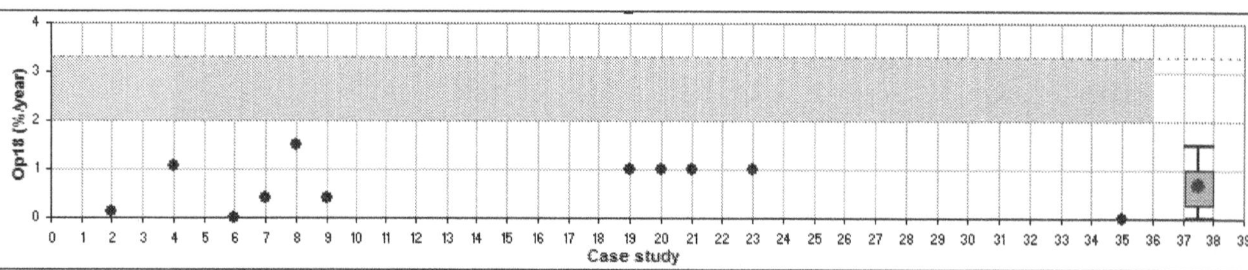

Comments: The mean value of the results collected is 0.68% of replaced valves per year and 50% are situated in the range from 0.26 to 1.5%. Most case studies present relatively low valve replacement rates, corresponding in average to a valve lifetime of 150 years, which is not realistic.

Difficulties: Several end-users had difficulties with D21 (Replaced or renewed mains) and with C29 (Mains valves). It seems that the balance between the cost of collecting these variables and the usefulness of the indicator may not be evident.

Relevance: This PI is considered important or useful almost by almost all end-users, and some recommend abandoning it.

PI guidance range: At this stage, a tentative recommended guidance range seems to be 2.0-3.3% of valves replaced per year, representing a global valve renovation within a period of 30 to 50 years.

Op19 - **SERVICE CONNECTION REHABILITATION** (%/year)	
Concept:	Number of service connections replaced or renovated during the year / total number of service connections x 100
Processing rule:	Op19 = D22/C32 x 100
Variables:	D22 – Service connection rehabilitation (No./year) C32 – Number of service connections (No.)

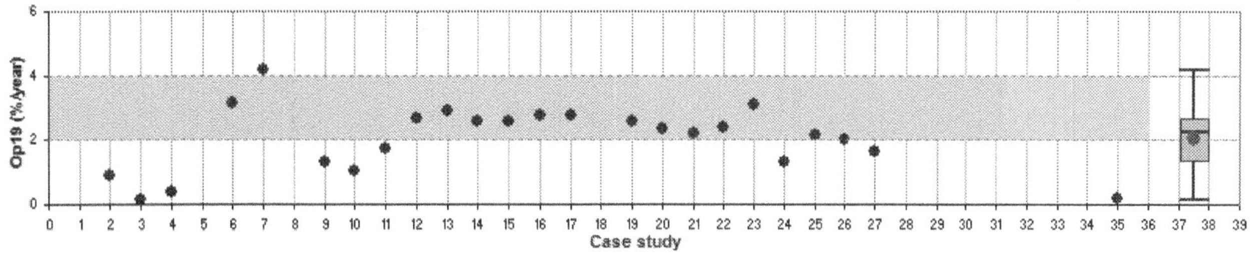

Comments:	The mean value of the results collected is 2.0% of service connections rehabilitated per year and 50% are situated in the range from 1.3 to 2.7%. Most case studies present values close to the sample mean, corresponding in average to a service connection lifetime of 50 years.
Difficulties:	Several end-users had difficulties with D22 (Service connection rehabilitation) and with C32 (Number of service connections).
Relevance:	This PI is considered essential or important by almost all end-users, and nobody recommends abandoning it.
PI guidance range:	At this stage, a tentative recommended guidance range seems to be 2.0-4.0% of service connections rehabilitated per year, representing a global service connection network renovation within a period of 25 to 50 years.

1.1.2 Failures and repairs

Op26 - **MAINS FAILURES** (No./100 km/year)	
Concept:	Number of mains failures during the year, including failures of pipes, valves, fittings and service connection insertion point failures / total mains length x 100
Processing rule:	Op26 = D25/C6 x 100
Variables:	D25 – Mains failures (No. / year) C6 – mains length (km)

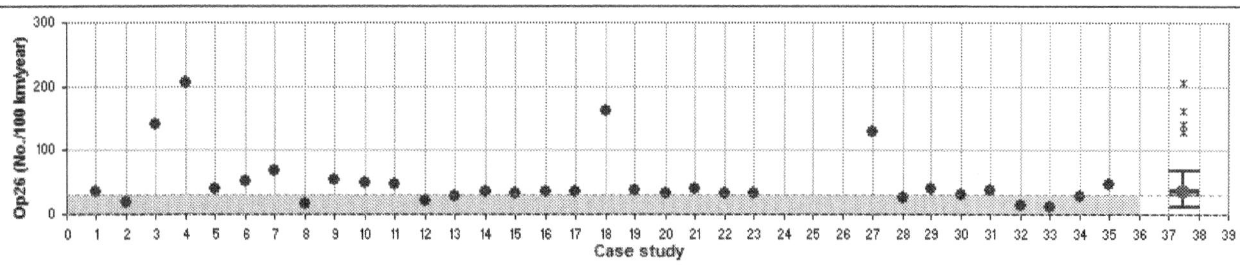

Comments:	The mean value of the results collected is 58 mains failures /100 km/year and 50% of the case studies are situated in the range from 30 to 51. Some case studies present a relatively high failure rate.
Difficulties:	No difficulties were reported with regard to this indicator.
Relevance:	This PI is considered essential by almost all end-users.
PI guidance range:	At this stage, a tentative recommended guidance range seems to be less or equal to 30 failures /100 km/year. These values are based on international statistics, and particularly on *Alegre & Dória (1998) – Develop decision criteria to prioritize replacement and rehabilitation of mains and appurtenances – survey results, American Water Works Research Foundation AWWARF 459, report 292/98 – NES, LNEC, Lisbon (49 pp.).*

Op26a - **PIPE FAILURES** (No./100 km/year)	
Concept:	Number of pipe failures during the year / total mains length x 100
Processing rule:	Op26a = D25a/C6 x 100
Variables:	D25a – pipe failures (No. /year) C6 – mains length (km)
Parent PI:	Op26

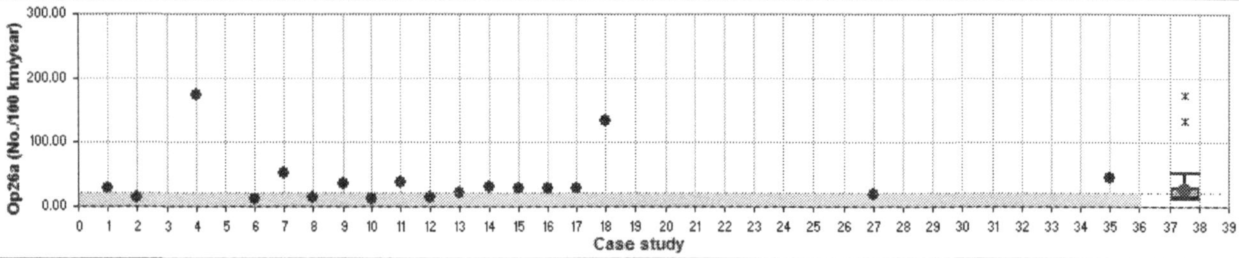

Comments:	The mean value of the results collected is 67 pipe failures /100 km/year and 50% of the case studies are situated in the range from 17 to 48. Some case studies present a relatively high failure rate. The fact that this PI is a subset of Op26 (Main failures) and has a higher mean value can be explained because the data samples do not coincide in both cases due to different number of responses.
Difficulties:	A few end-users had difficulties with D25a (Pipe failures), informing that this information is not recorded separately from other types mains failures.
Relevance:	This PI is considered essential by almost all end-users.
PI guidance range:	At this stage, a tentative recommended guidance range seems to be less or equal to 20 failures /100 km/year.

Op26b - **JOINT FAILURES** (No./100 km/year)	
Concept:	Number of joint failures during the year / total mains length x 100
Processing rule:	Op26b = D25b/C6 x 100
Variables:	D25b – Joint failures (No./year) C6 – Mains length (km)
Parent PI:	Op26

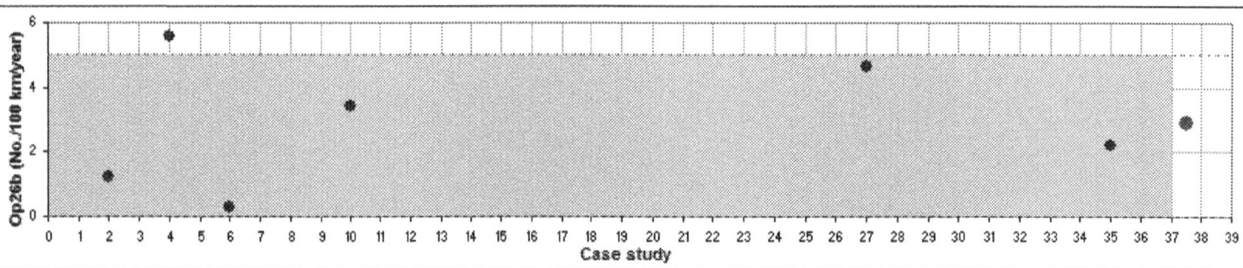

Comments:	Only 5 results were provided, in a range from 0.30 to 5.6 joint failures per 100 km/year. With such a small sample, no trends can be identified.
Difficulties:	Most end-users had difficulties with D25b (Joint failures), informing that this information is not recorded separately from other types mains failures, and one end-user recommends abandoning this indicator.
Relevance:	This PI is considered essential, important or useful by almost all end-users.
PI guidance range:	At this stage, a tentative recommended guidance range seems to be less or equal to 5 failures /100 km/year.

Op26c - **VALVE FAILURES** (No./100 km/year)	
Concept:	Number of valve failures during the year / total mains length x 100
Processing rule:	Op26c = D25c/C6 x 100
Variables:	D25c – Valve failures (No) C6 – Mains length (km)

Parent PI:	Op26

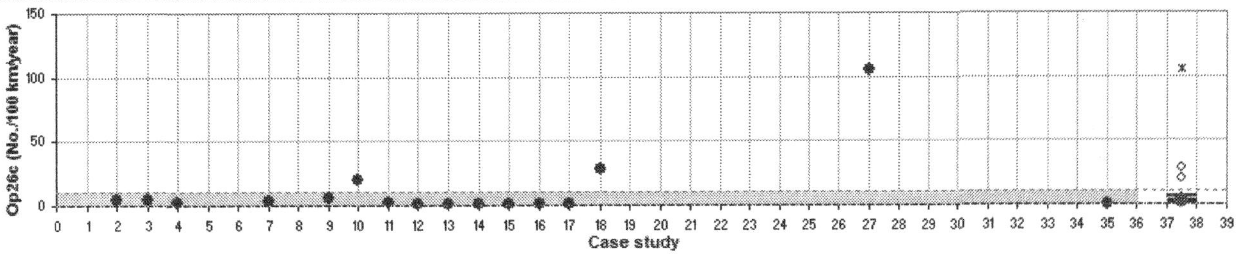

Comments:	The mean value of the results collected is 13 valve failures /100 km/year and 50% of the case studies are situated in the range from 0.75 to 4.8. Some case studies present a relatively high failure rate. The fact that this PI is a subset of Op26 (Main failures) and has a higher mean value can be explained because the data samples do not coincide in both cases due to different number of responses. One case study presents a much higher result.
Difficulties:	Several end-users had difficulties with D25c (Valve failures), informing that this information is not recorded separately from other types of mains failures.
Relevance:	This PI is considered important or useful by most end-users.
PI guidance range:	At this stage, a tentative recommended guidance range seems to be less or equal to 10 failures /100 km/year.

Op26d - **SERVICE CONNECTION INSERTION POINT FAILURES** (No./100 km/year)

Concept:	Number of failures that occur in the insertion point of the service connection during the year / total mains length x 100
Processing rule:	Op26d = D25d/C6 x 1000
Variables:	D25d – Service connection insertion point failures
	C6 – mains length (km)
Parent PI:	Op26

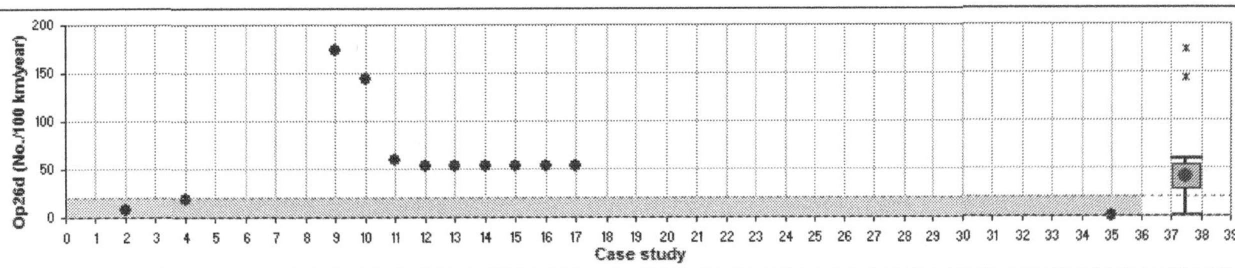

Comments:	The mean value of the results collected is 60 service connection insertion point failures /100 km/year and 50% of the case studies are situated in the range from 45 to 55. Some case studies present a relatively high failure rate. The fact that this PI is a subset of Op26 (Main failures) and has a higher mean value can be explained because the data samples do not coincide in both cases due to different number of responses.
Difficulties:	Most end-users had difficulties with D25d (Service connection insertion point failures), informing that this information is not recorded separately from other types of mains failures.
Relevance:	This PI is considered essential, important or useful by almost all end-users.
PI guidance range:	At this stage, a tentative recommended guidance range seems to be less or equal to 20 failures /100 km/year.

Op26e - **CRITICAL MAINS FAILURES** (No./100 km/year)

Concept:	Number of mains failures in sensitive areas during the year / total mains length in sensitive areas x 100
Processing rule:	Op26e = D25f/C6a x 100
Variables:	D25f – Critical mains failures
	C6a – Mains length in sensitive areas

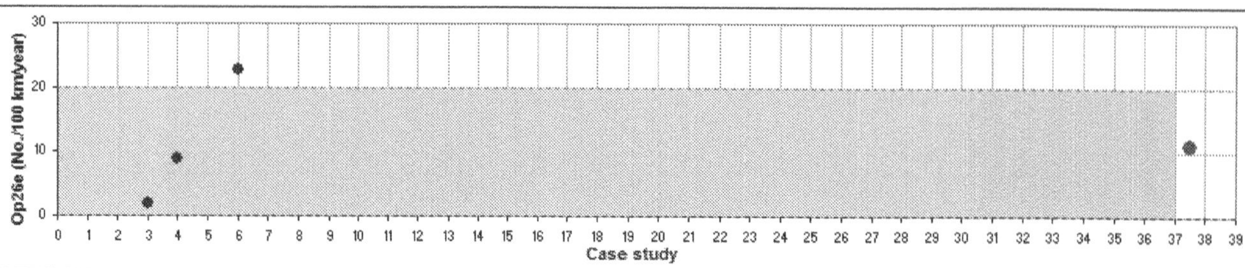

Comments: Only 3 results were provided, in a range of 1.9 to 23 critical mains failures /100 km/year. With such a small sample, no trends can be identified.

Difficulties: Most end-users had difficulties with the interpretation on sensitive areas and therefore with the assessment of this indicator. Most end-users had difficulties with D25f (Critical mains failures) and with C6a (Mains length in sensitive areas), informing that this information is not recorded separately. Several end-users recommend abandoning this indicator.

Relevance: This PI is considered essential or important by most end-users.

PI guidance range: At this stage, a tentative recommended guidance range seems to be less or equal to 5 failures /100 km/year.

Op27 - **SERVICE CONNECTION FAILURES** (No./1000 connections/year)

Concept: Number of service connection failures during the year / number of service connections x 1000

Processing rule: Op27 = D26/C32 x 1000

Variables: D26 – Service connection failures

C32 – Number of service connections

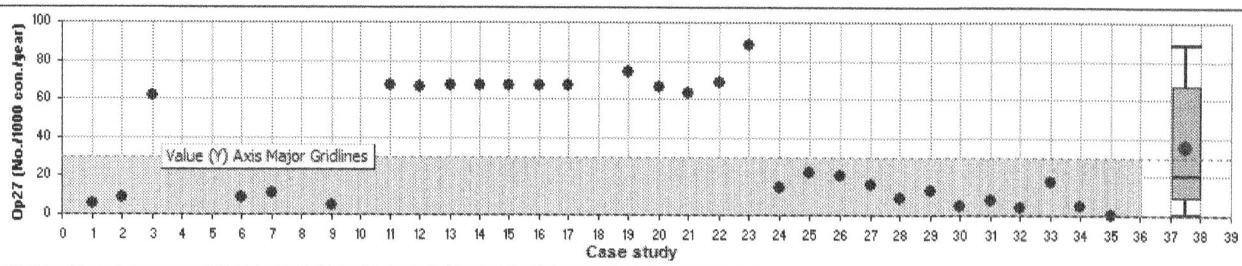

Comments: The mean value of the results collected is 36 service connection failures /100 km/year and 50% of the case studies are situated in the range from 8.4 to 67. Some case studies present a relatively high failure rate. The fact that this PI is a subset of Op26 (Main failures) and has a higher mean value can be explained because the data samples do not coincide in both cases due to different number of responses.

Difficulties: Several end-users had difficulties with D26 (Serviced connection failures) and one with C32 (Number of service connections).

Relevance: This PI is considered essential or important by almost all end-users.

PI guidance range: At this stage, a tentative recommended guidance range seems to be less or equal to 30 failures /1000 connections/year.

Op28 - **HYDRANT FAILURES** (No./1000 hydrants/year)

Concept: Number of hydrant failures during the year / total number of hydrants x 1000

Processing rule: Op28 = D27/C31 x 1000

Variables: D27 – Hydrant failures

C31 – Hydrants

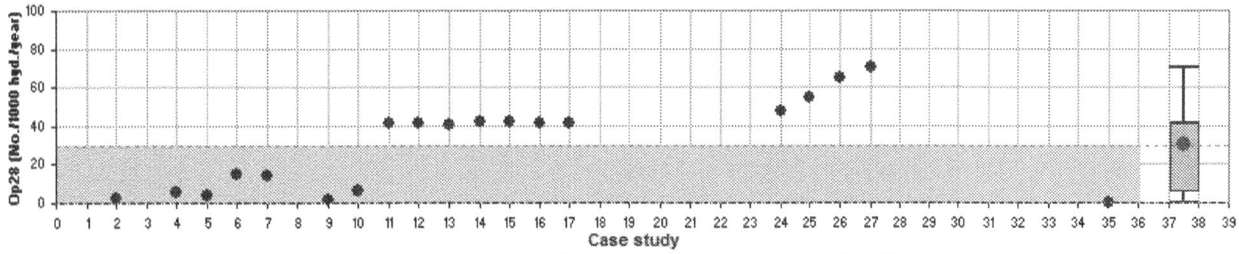

Comments:	The mean value of the results collected is 30 hydrant failures /1000 hydrants/year and 50% of the case studies are situated in the range from 5.6 to 42. Some case studies present a relatively high failure rate.
Difficulties:	Several end-users had difficulties with D27 (Hydrant failures) and one with C31 (Hydrants). One end-user recommends abandoning this indicator.
Relevance:	This PI is considered important by most end-users.
PI guidance range:	At this stage, a tentative recommended guidance range seems to be less or equal to 30 failures /1000 hydrants/year.

Op29 - **POWER FAILURES** (hours/pumping station/year)

Concept:	Sum, for all pumps, of the number of hours each pumping station is out of service due to power supply interruptions during year / total number of pumping stations
Processing rule:	Op29 = D28/C4
Variables:	D28 – Power failures C4 – Pumping stations

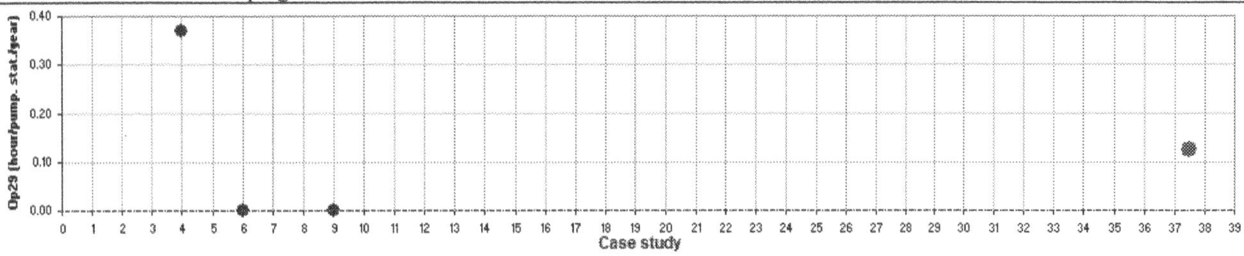

Comments:	Only 3 results were provided, in a range of 0.0 to 0.37 hours/pumping station/year. With such a small sample, no trends can be identified.
Difficulties:	Most end-users had difficulties with D28 (Power failures), informing that this information is not recorded. Three end-users recommend abandoning this indicator.
Relevance:	This PI is considered useful, irrelevant or not applicable by almost all end-users.
PI guidance range:	At this stage, there is not enough information for the definition of a recommended threshold.

Op5 - ACTIVE LEAKAGE CONTROL REPAIRS (No./100 km/year)

Concept: Number of main leaks detected and repaired due to active leakage control / total mains length x 100

Processing rule: Op5 = D9/C6 x 100

Variables: D9 – Leakage repair due to active leakage control

C6 – Mains length (km)

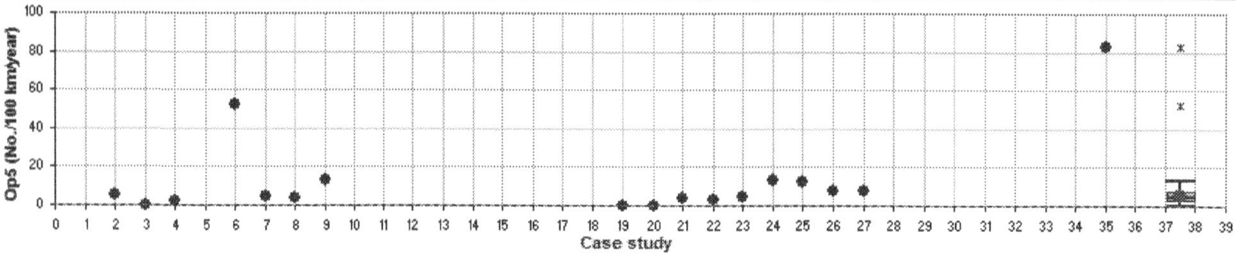

Comments: The mean value of the results collected is 13 repairs due to active leakage control / 100 km/year and 50% of the case studies are situated in the range from 2.9 to 13.

Difficulties: Several end-users had difficulties with D9 (Leakage repair due to active leakage control), informing that this information is not recorded. One end-user recommends abandoning this indicator.

Relevance: This PI is considered important or useful by most end-users.

PI guidance range: It does not seem to be advisable to establish a guidance range for this indicator because it depends much from case to case.

1.1.3 Water losses

Op22 - WATER LOSSES PER CONNECTION (m^3/connection/year)

Concept: Water losses / number of service connections

Processing rule: Op22 = A20/C32

Variables: A20 – Water losses

C32 – Number of service connections

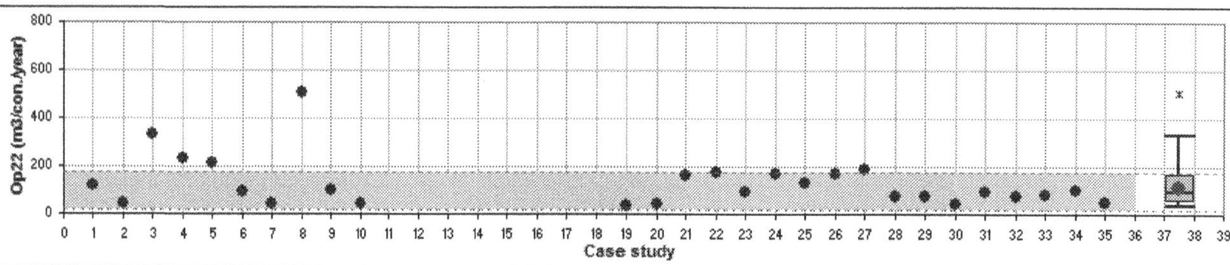

Comments: The mean value of the results collected is 153 m^3 /connection/year and 50% of the case studies are situated in the range from 75 to 172. Some case studies present relatively higher results.

Difficulties: A few end-users had difficulties with A20 (Water losses) and one with C32 (Number of service connections), informing that this information is not recorded. Two end-users recommend abandoning this indicator.

Relevance: This PI is considered essential or important by almost all end-users.

PI guidance range: The tentative recommended guidance range was set between 20 to 175 m^3/connection/year. 20 m^3/connection/year corresponds to a system with very little apparent losses and real losses of 40 l/connection/day. 175 m^3/connection/year corresponds to a network with 300 l/connection/day of real losses and 60% of this value for apparent losses (see Op24). These values were partially based on *IWA (2000) – "Blue pages on losses from water supply systems", ed. A. Lambert e W. Hirner, International Water Association.*

Op24 - REAL LOSSES PER CONNECTION (l/connection/day when system is pressurised)

Concept: Real losses x 1000 / (number of service connections x 365 x T/100) (T = % of year system is pressurised)

Processing rule: Op24 = A24 x 1000/(C32x365xD29/365/24)

Variables: A24 – Real losses (m3/year)

C32 – Number of service connections (No.)

D29 – Time system is pressurised (hours/year)

Parent PI: Op22

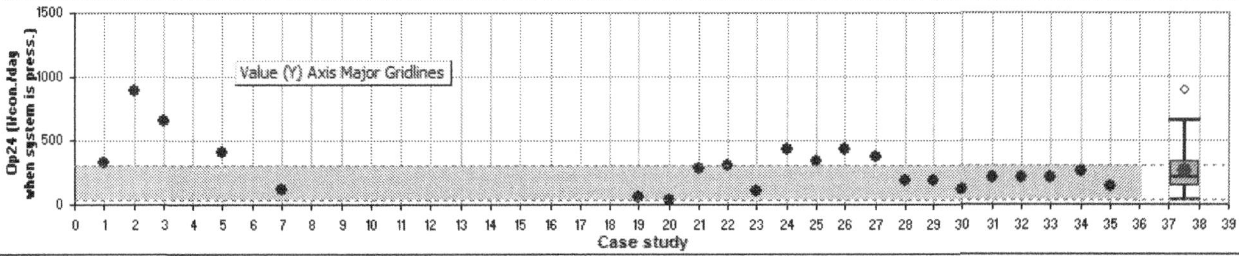

Comments: The mean value of the results collected is 308 l/connection/day when system is pressurised and 50% of the case studies are situated in the range 143 to 381. One case study presents a relatively higher result.

Difficulties: Most end-users had difficulties with A24 (Real losses), and a few with C32 (Number of service connections). Three end-users recommend abandoning this indicator. Results seem to show that some end-users misinterpreted the meaning and processing rule of this indicator.

Relevance: This PI is considered important by several end-users, and essential, irrelevant or not applicable in equal parts by the remaining.

PI guidance range: The tentative recommended guidance range was set between 40 to 300 l/connection/day. According to the IWA Water Losses group, 40 may correspond to a very good network with about 30 service connections per km, a service pressure of 300 kPa and very short service connections; 300 l/connection/day may correspond to 3 times the average real losses of a very good network with a service connection density of 30 /km, a service pressure of 600 kPa and connections of about 20 m long. These values were partially based on IWA (2000) – "Blue pages on losses from water supply systems", ed. A. Lambert e W. Hirner, International Water Association.

Op25 - INFRASTRUCTURE LEAKAGE INDEX (-)

Concept: Real losses (Op24)/ technical achievable low-level annual real losses (when system is pressurised)

Processing rule: Op25 = Op24/(18xC6/C32+0.7+0.025xC33)/D31

Variables: A24 – Real losses

C32 – Number of service connections (No.)

D29 – Time system is pressurised (hours/year)

C6 – Mains length (km)

C33 – Average service connections length to measurement point (m)

D31 – Average operating pressure (m)

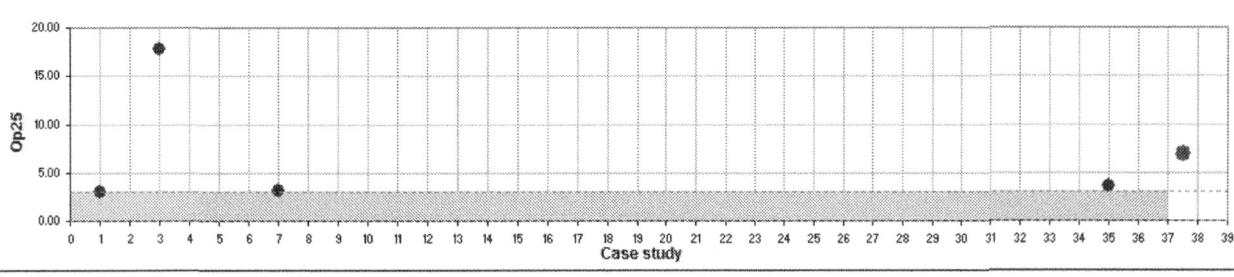

Comments: Only 4 results were provided, in a range of 3.0 to 18. With such a small sample, no trends can be identified.

Difficulties: Most end-users had difficulties with at one or more variables. Four end-users recommend abandoning this indicator.

Relevance: This PI is considered useful by several end-users, and important or irrelevant by the remaining.

PI guidance range: At this stage, a tentative recommended guidance range seems to be less or equal 3. According to the IWA Water Losses group, well-managed systems are expected to have a leakage index close to 1. These values were partially based on IWA (2000) – "Blue pages on losses from water supply systems", ed. A. Lambert e W. Hirner, International Water Association.

1.2 Quality of service indicators

1.2.1 Service

QS9 - PRESSURE OF SUPPLY ADEQUACY (%)

Concept:	Number of delivery points that receive and are likely to receive pressure equal to or above the guaranteed or declared target level at the peak demand hour (but not when demand is abnormal) / number service connections x 100
Processing rule:	QS9 = D30/C32 x 100
Variables:	D30 – Delivery points with adequate pressure C32 – Number of service connections

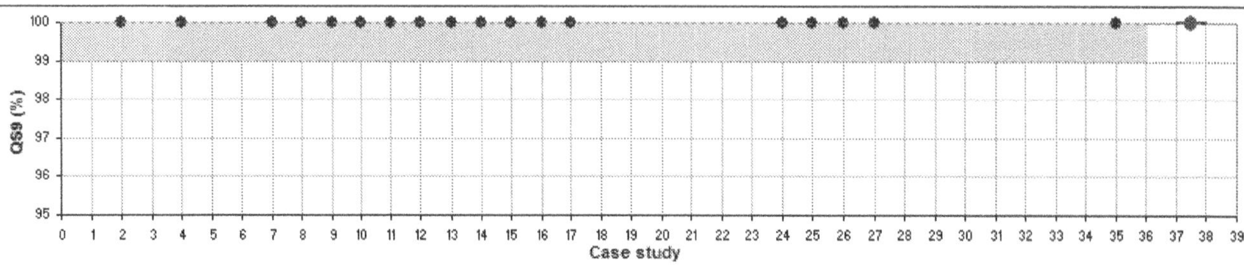

Comments:	All the results collected are equal to 100%.
Difficulties:	Most end-users had difficulties with D30 (Delivery points with adequate pressure), and a few with C32 (Number of service connections). One end-user recommends abandoning this indicator.
Relevance:	This PI is considered important or useful by most end-users.
PI guidance range:	At this stage, a tentative recommended guidance range seems to be more than 99%.

QS11 - WATER INTERRUPTIONS (%)

Concept:	Σ (Population subject to a water interruption x duration of the interruption in hours) / (population served x 24 x 365) x 100
Processing rule:	QS11 = D32/(F1x24x365) x 100
Variables:	D32 – Water interruptions (No.) F1 – Population supplied (No.)

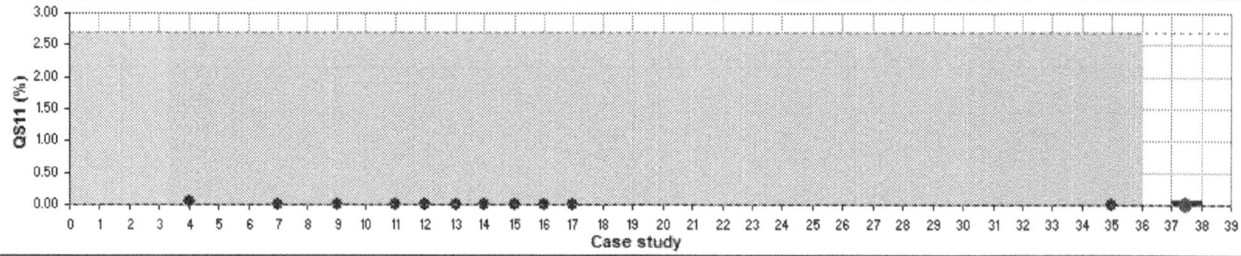

Comments:	The mean value of the results collected is 0% and 50% of the case studies are also equal to 0.
Difficulties:	Most end-users had difficulties with D32 (Water interruptions), informing that this information is not recorded. Two end-users recommend abandoning this indicator.
Relevance:	This PI is considered useful by most end-users.
PI guidance range:	At this stage, a tentative recommended guidance range seems to be less or equal 2.7%, corresponding to an average of 1 unplanned interruption per connection, with a duration of at least 12 h, every 5 years.

QS12 - **INTERRUPTIONS PER CONNECTION** (No./1000 connections)	
Concept:	Number of interruptions / number of service connections x 1000
Processing rule:	QS12 = D33/C32 x 1000
Variables:	D33 – Service interruptions (No.) C32 – Number of service connections (No.)

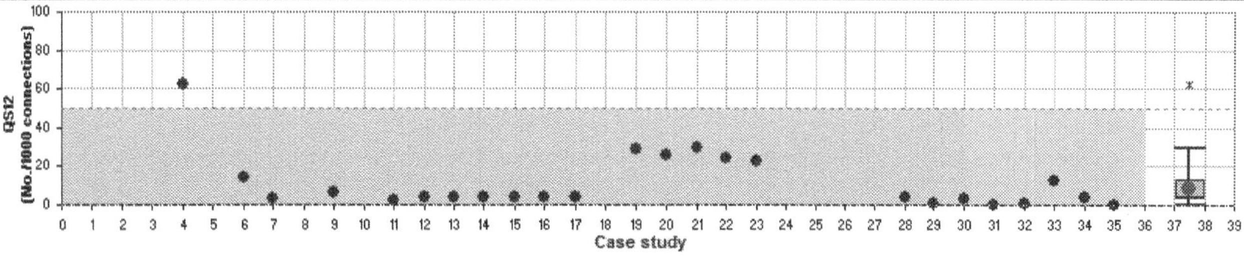

Comments:	The mean value of the results collected is 12 interruptions per 1000 service connections and 50% of the case studies are situated in the range 3.3 to 18.
Difficulties:	Several end-users had difficulties with D33 (Service interruptions) and one with C32 (Number of service connections). One end-user recommends abandoning this indicator.
Relevance:	This PI is considered useful by almost all end-users.
PI guidance range:	At this stage, a tentative recommended guidance range seems to be less or equal 50, corresponding to an average of 1 unplanned interruption per connection, with duration of at least 12 h, every 20 years.

QS12a - **CRITICAL INTERRUPTIONS PER CONNECTION** (No./1000 connections)	
Concept:	Number of supply interruptions to sensitive service connections during the year / number of sensitive service connections x 1000
Processing rule:	QS12a = D33a/C32a x 1000
Variables:	D33a – Critical interruptions (No.) C32a – Number of sensitive service connections (No.)
Parent PI:	QS12

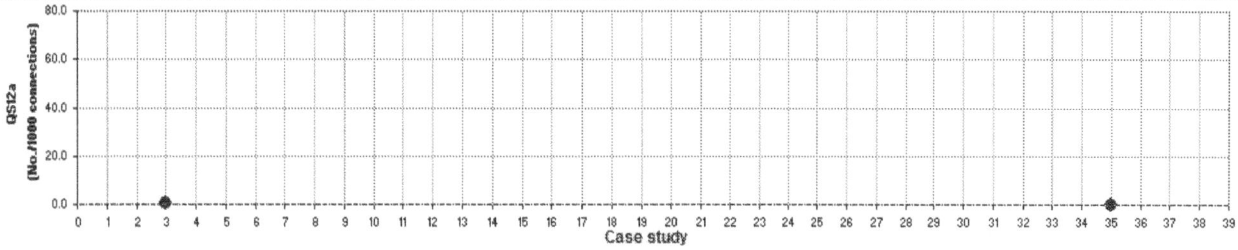

Comments:	Only 4 results were provided within a range from 0 to 800 critical interruptions per 1000 connections. Therefore no trends can be identified.
Difficulties:	Most end-users had difficulties with the interpretation of critical interruption and therefore with the assessment of this indicator. Several end-users had difficulties with D32 (Critical interruptions) and with C32a (Number of sensitive service connections). Three end-users recommend abandoning this indicator.
Relevance:	This PI is considered useful by almost all end-users.
PI guidance range:	It does not seem to be advisable to establish a guidance range for this indicator with the available information.

QS13 - POPULATION EXPERIENCING RESTRICTIONS TO WATER SERVICE (%)

Concept:	Σ (Population affected by restrictions to water service x duration of the restrictions to water service in hours) / (total population served x 24 x 365) x 100
Processing rule:	QS13 = D34/(F1x24x365) x 100
Variables:	D34 – Water use restrictions (hours)
	F1 – Population supplied (No.)

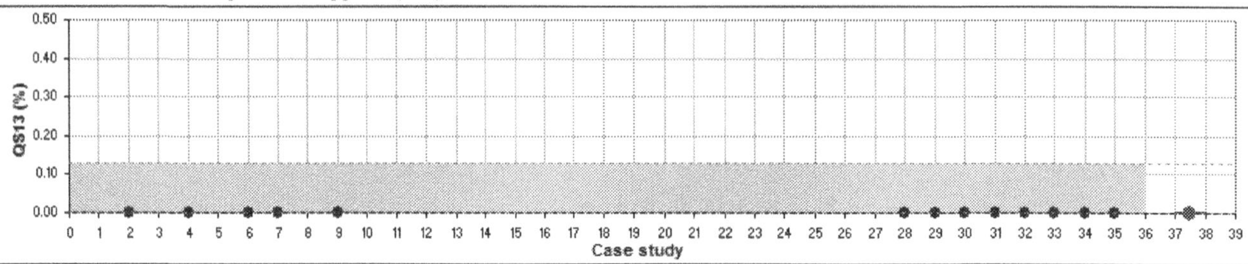

Comments:	The mean value of the results collected is 0% and 50% of the case studies are also equal to 0.
Difficulties:	Several end-users had difficulties with D34 (Water use restrictions) and one with F1 (Population supplied). Four end-users recommend abandoning this indicator.
Relevance:	This PI is considered irrelevant by most end-users.
PI guidance range:	At this stage, a tentative recommended guidance range seems to be less or equal 0.13%, corresponding approximately to an average of 1 week of restrictions to all the population supplied every 15 years.

QS14 - DAYS WITH RESTRICTIONS TO WATER SERVICE (%)

Concept:	Total number of days with restrictions to water service during the year / 365 x 100
Processing rule:	QS14 = D35/365 x 100
Variables:	D35 – Number of days with restrictions to water service (days)

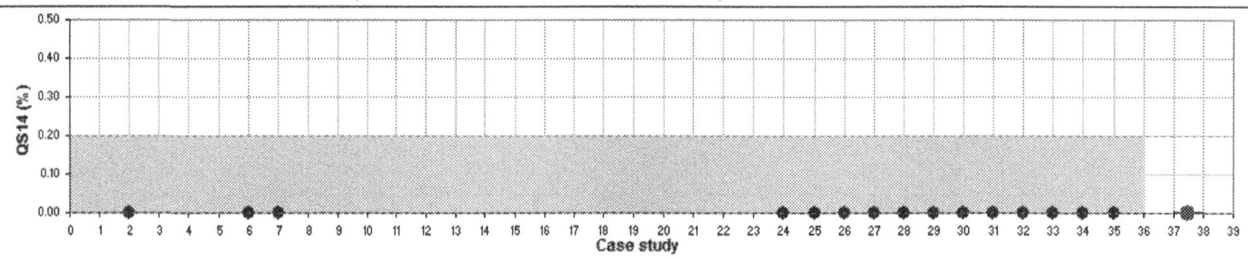

Comments:	The mean value of the results collected is 0% and 50% of the case studies are also equal to 0.
Difficulties:	Several end-users had difficulties with D35 (Number of days with restrictions to water service), informing that this information is not recorded. One end-user recommends abandoning this indicator.
Relevance:	This PI is considered important or useful by most end-users.
PI guidance range:	At this stage, a tentative recommended guidance range seems to be less or equal 0.2%, corresponding to an average of 1 week of restrictions every 10 years.

QS15 - **QUALITY OF SUPPLIED WATER (%)**	
Concept:	Total number of treated water tests complying with the applicable standards or legislation during the year / total number of tests of treated water performed during the year x 100
Processing rule:	QS15 = (D51+D52+D53)/D41 x 100
Variables:	D51 – Compliance of aesthetic tests (No.) D52 – Compliance of microbiological tests (No.) D53 – Compliance of physical-chemical tests (No.) D41 – Water quality tests performed (No.)

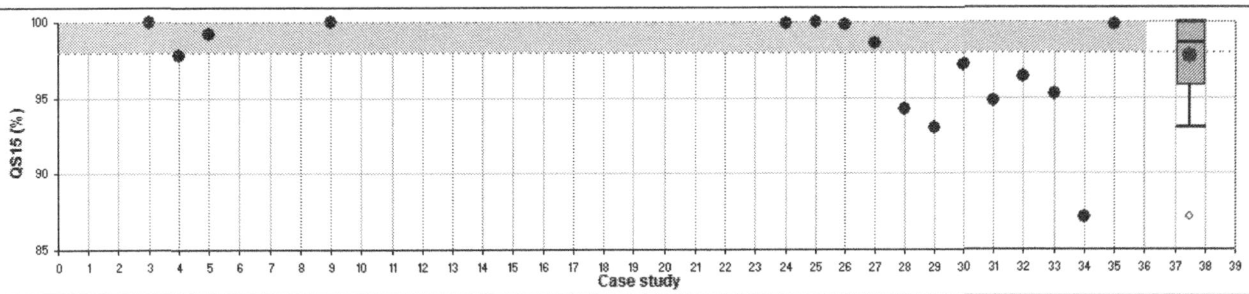

Comments:	The mean value of the results collected is 96% and 50% of the case studies are situated in the range 95 to 100%.
Difficulties:	Several end-users had difficulties with D51 (Compliance of aesthetic tests), D52 (Compliance of microbiological tests) and D53 (Compliance of physical-chemical tests), informing that these data are not recorded. No end-users recommend abandoning this indicator.
Relevance:	This PI is considered essential or important most end-users.
PI guidance range:	At this stage, a tentative recommended guidance range seems to be greater than 98%.

QS16 - **AESTHETIC TESTS COMPLIANCE (%)**	
Concept:	Number of aesthetic tests of treated water complying with the applicable standards or legislation during the year / total number of aesthetic tests treated water performed during the year x 100
Processing rule:	QS16 = D51/D42 x 100
Variables:	D51 – Compliance of aesthetic tests (No.) D42 – Aesthetic tests performed (No.)
Parent PI:	QS15

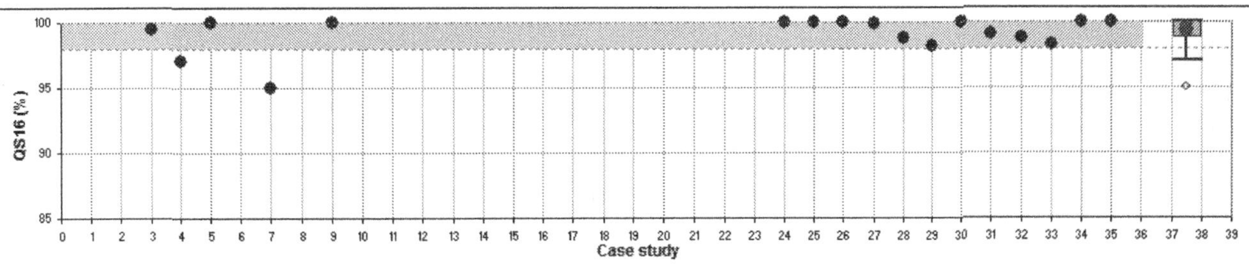

Comments:	The mean value of the results collected is 99% and 50% of the case studies are situated in the range 98 to 100%.
Difficulties:	Several end-users had difficulties with D51 (Compliance of aesthetic tests) and D42 (Aesthetic tests performed), informing that these data are not recorded. No end-users recommend abandoning this indicator.
Relevance:	This PI is considered essential or important by most end-users.
PI guidance range:	At this stage, a tentative recommended guidance range seems to be greater than 98%.

QS16a - WATER TASTE TESTS COMPLIANCE (%)

Concept:	Number of water taste tests of treated water complying with the applicable standards or legislation during the year / total number of water taste tests of treated water performed during the year x 100
Processing rule:	QS16a = D51a/D42a x 100
Variables:	D51a – Compliance of water taste tests (No.)
	D42a – Water taste tests performed (No.)
Parent PI:	QS16

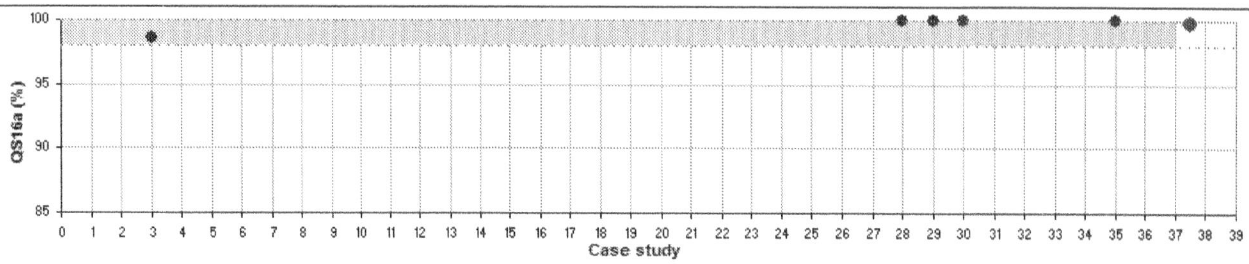

Comments:	Only 3 results were provided, in a range of 99 to 100%. With such a small sample, no trends can be identified.
Difficulties:	Most end-users had difficulties with D51a (Compliance of water taste tests) and D42a (Water taste tests performed), informing that these data are not recorded. Two end-users recommend abandoning this indicator.
Relevance:	This PI is considered essential or important by most end-users.
PI guidance range:	At this stage, a tentative recommended guidance range seems to be greater than 98%.

QS16b - WATER COLOUR TESTS COMPLIANCE (%)

Concept:	Number of water colour tests of treated water complying with the applicable standards or legislation during the year / total number of water colour tests of treated water performed during the year x 100
Processing rule:	QS16 b = D51b/D42b x 100
Variables:	D51b – Compliance of water colour tests (No.)
	D42b – Water colour tests performed (No.)
Parent PI:	QS16

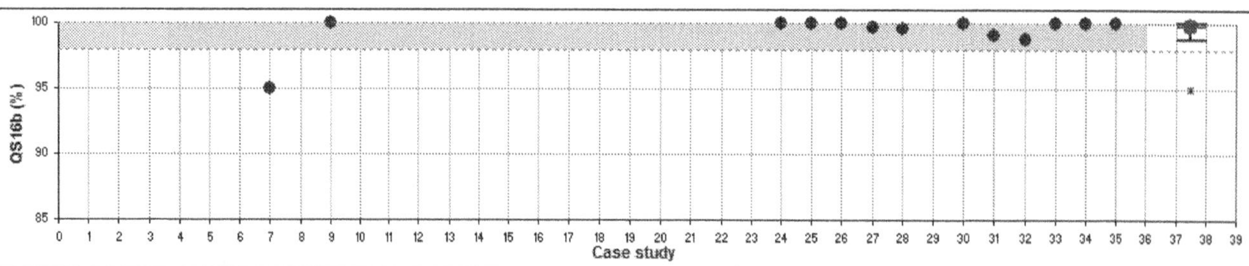

Comments:	The mean value of the results collected is 99% and 50% of the case studies are equal to 100%.
Difficulties:	Several end-users had difficulties with D51b (Compliance of water colour tests) and D42b(Water colour tests performed), informing that these data are not recorded. One end-user recommends abandoning this indicator.
Relevance:	This PI is considered essential or important by most end-users.
PI guidance range:	At this stage, a tentative recommended guidance range seems to be greater than 98%.

QS17 - MICROBIOLOGICAL TESTS COMPLIANCE (%)	
Concept:	Number of microbiological tests of treated water complying with the applicable standards or legislation during the year / total number of microbiological tests of treated water performed during the year x 100
Processing rule:	QS17 = D52/D43 x 100
Variables:	D52 – Compliance of microbiological tests (No.) D43 – Microbiological tests performed (No.)
Parent PI:	QS15

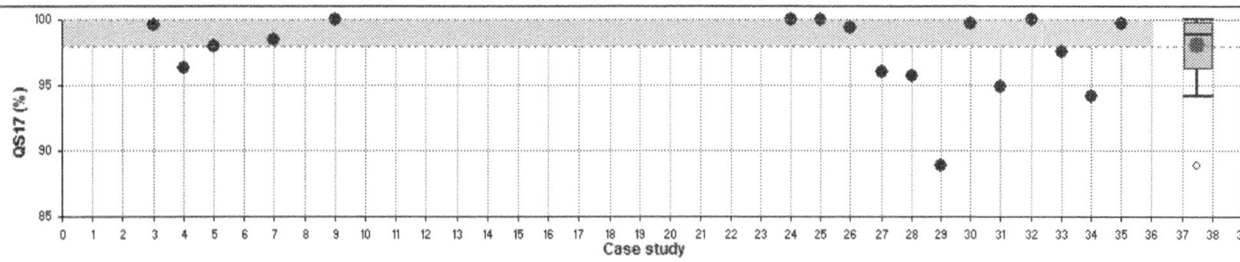

Comments:	The mean value of the results collected is 98% and 50% of the case studies are situated in the range 96 to 100%.
Difficulties:	A few end-users had difficulties with D52 (Compliance of microbiological tests) and D43 (Microbiological tests performed), informing that these data are not recorded. No end-user recommends abandoning this indicator.
Relevance:	This PI is considered essential or important by all end-users.
PI guidance range:	At this stage, a tentative recommended guidance range seems to be greater than 98%.

QS18 - PHYSICAL-CHEMICAL TESTS COMPLIANCE (%)	
Concept:	Number of physical-chemical tests of treated water complying with the applicable standards or legislation during the year / total number of physical-chemical tests of treated water performed during the year x 100
Processing rule:	QS18 = D53/D44 x 100
Variables:	D53 – Compliance of water colour tests (No.) D44 – Water colour tests performed (No.)
Parent PI:	QS15

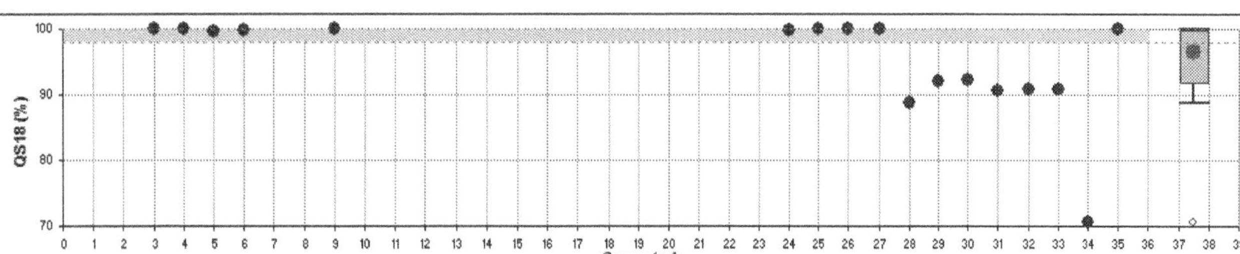

Comments:	The mean value of the results collected is 96% and 50% of the case studies are situated in the range 91 to 100%.
Difficulties:	Several end-users had difficulties with D53 (Compliance of water colour tests) and D44 (Water colour tests performed), informing that these data are not recorded. No end-user recommends abandoning this indicator.
Relevance:	This PI is considered essential or important by all end-users.
PI guidance range:	At this stage, a tentative recommended guidance range seems to be greater than 99%.

1.2.2 Customer complaints

QS22 - SERVICE COMPLAINTS PER CONNECTION (No. complaints/1000 connections/year)

Concept: Number of service complaints during the year / number of service connections x 1000

Processing rule: QS22 = F11/C32 x 1000

Variables: F11 – Service customer complaints (No.)

C32 – Number of service connections (No.)

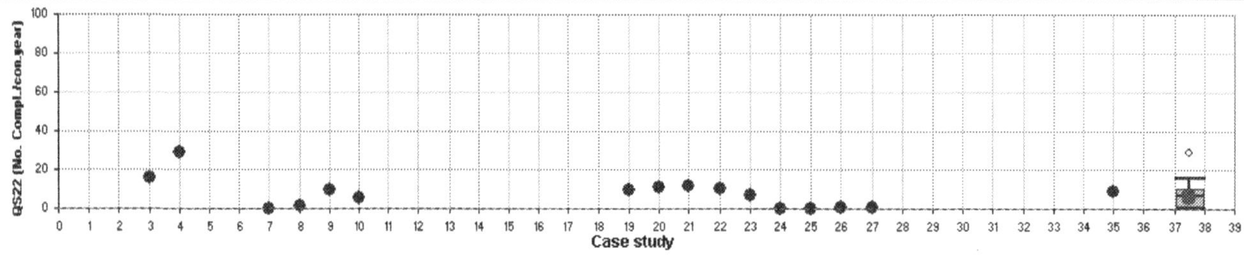

Comments: The mean value of the results collected is 13 service complaints/1000 connections/year and 50% of the case studies are situated in the range from 0.37 to 11. One case study presents a relatively higher result.

Difficulties: Several end-users had difficulties with F11 (Service customer complaints) and C32 (Number of service connections), informing that this information is not recorded. Two end-users recommend abandoning this indicator.

Relevance: This PI is considered important or useful by almost all end-users.

PI guidance range: The number of complaints depends much on cultural factors and therefore it does not seem advisable to establish a guidance range for this indicator.

QS23 - PRESSURE COMPLAINTS (%)

Concept: Number of pressure complaints during the year / number of service complaints during the year x 100

Processing rule: QS23 = F12/F11 x 100

Variables: F12 – Pressure customer complaints (No.)

F11 – Service customer complaints (No.)

Parent PI: QS22

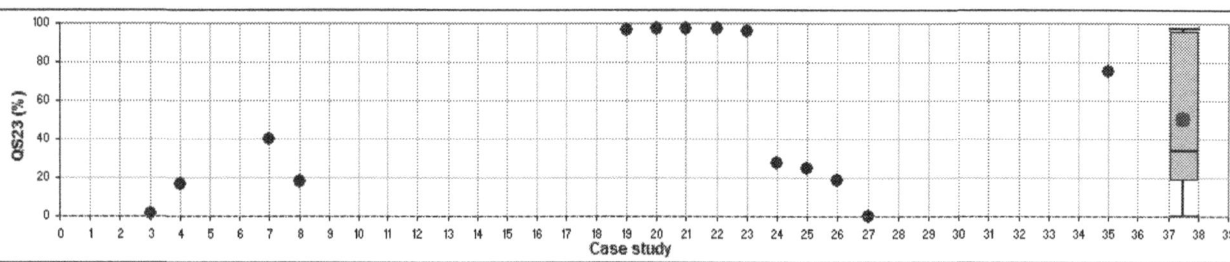

Comments: The mean value of the results collected is 46% and 50% of the case studies are situated in the range from 17 to 96%.

Difficulties: Most end-users had difficulties with F11 (Service customer complaints) and F12 (Pressure customer complaints), informing that these data are not recorded. Two end-users recommend abandoning this indicator.

Relevance: This PI is considered important or useful by almost all end-users.

PI guidance range: The number of complaints depends much on cultural factors and therefore it does not seem advisable to establish a guidance range for this indicator.

QS24 - CONTINUITY COMPLAINTS (%)

Concept:	Number of continuity complaints during the year / number of service complaints during the year x 100
Processing rule:	QS24 = F13/F11 x 100
Variables:	F13 – Continuity customer complaints (No.)
	F11 – Service customer complaints (No.)
Parent PI:	QS22

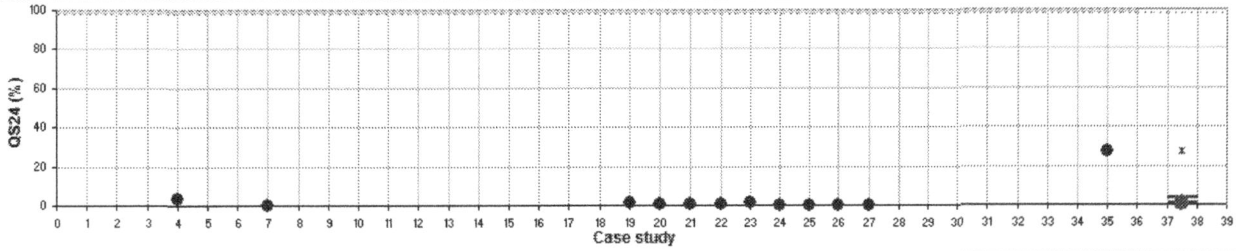

Comments:	The mean value of the results collected is 0.97% and 50% of the case studies are situated in the range 0 to 1.3%.
Difficulties:	Most end-users had difficulties with F13 (Continuity customer complaints) and F11 (Service customer complaints), informing that these data are not recorded. Two end-users recommend abandoning this indicator.
Relevance:	This PI is considered important or useful by most end-users.
PI guidance range:	The number of complaints depends much on cultural factors and therefore it does not seem advisable to establish a guidance range for this indicator.

QS25 - WATER QUALITY COMPLAINTS (%)

Concept:	Number of water quality complaints during the year / number of service complaints during the year x 100
Processing rule:	QS25 = F14/F11 x 100
Variables:	F14 – Service customer complaints (No.)
	F11 – Water quality customer complaints (No.)
Parent PI:	QS22

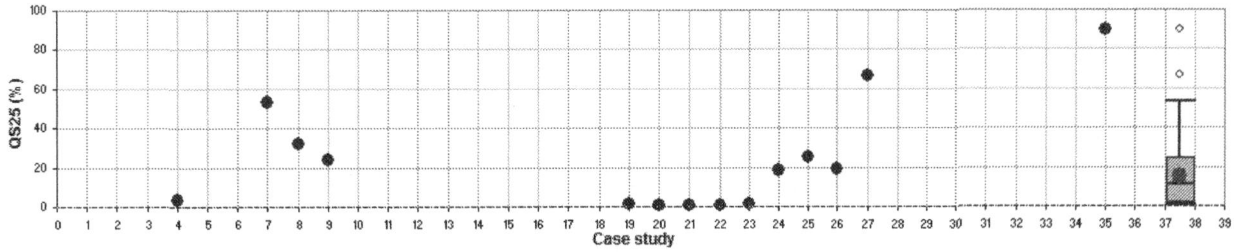

Comments:	The mean value of the results collected is 18 and 50% of the case studies are situated in the range 1.3 to 25%.
Difficulties:	Most end-users had difficulties with F14 (Water quality customer complaints) and F11 (Service customer complaints), informing that these data are not recorded. Two end-users recommend abandoning this indicator.
Relevance:	This PI is considered important by most end-users.
PI guidance range:	The number of complaints depends much on cultural factors and therefore it does not seem advisable to establish a guidance range for this indicator.

QS25a - WATER TASTE COMPLAINS (%)	
Concept:	Number of water taste complaints during the year / number of service complaints during the year x 100
Processing rule:	QS25a = F14a/F11 x 100
Variables:	F14a – Water taste customer complaints (No.) F11 – Service customer complaints (No.)
Parent PI:	QS25

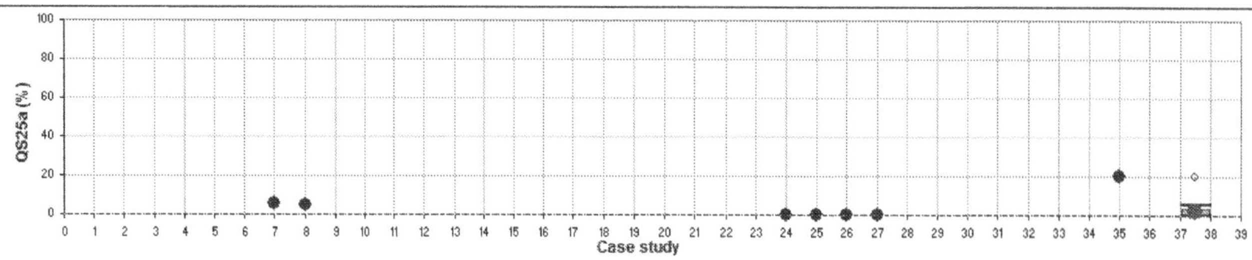

Comments:	Only 7 results were provided, in a range of 0.0 to 5.8%. With such a small sample, no trends can be identified.
Difficulties:	Most end-users had difficulties with F14a (Water taste customer complaints) and F11 (Service customer complaints), informing that these data are not recorded. Two end-users recommend abandoning this indicator.
Relevance:	This PI is considered important or useful by most end-users.
PI guidance range:	The number of complaints depends much on cultural factors and therefore it does not seem advisable to establish a guidance range for this indicator.

QS25b - WATER COLOUR COMPLAINS (%)	
Concept:	Number of water colour complaints during the year / number of service complaints during the year x 100
Processing rule:	QS25b = F14b/F11 x 100
Variables:	F14b – Water colour customer complaints (No.) F11 – Service customer complaints (No.)
Parent PI:	QS25

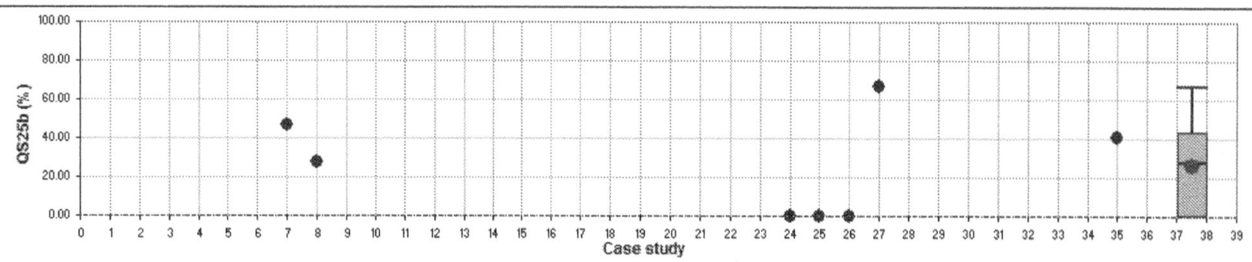

Comments:	The mean value of the results collected is 18% and 50% of the case studies are situated in the range 1.3 to 25%.
Difficulties:	Most end-users had difficulties with F14b (Water colour customer complaints) and F11 (Service customer complaints), informing that these data are not recorded. Two end-users recommend abandoning this indicator.
Relevance:	This PI is considered important or useful by almost all end-users.
PI guidance range:	The number of complaints depends much on cultural factors and therefore it does not seem advisable to establish a guidance range for this indicator.

QS26 - INTERRUPTIONS COMPLAINS (%)

Concept: Number of complaints due to supply interruptions during the year / number of service complaints during the year x 100

Processing rule: QS26 = F15/F11 x 100

Variables: F15 – Customer complaints on interruptions (No.)

F11 – Service customer complaints (No.)

Parent PI: QS22

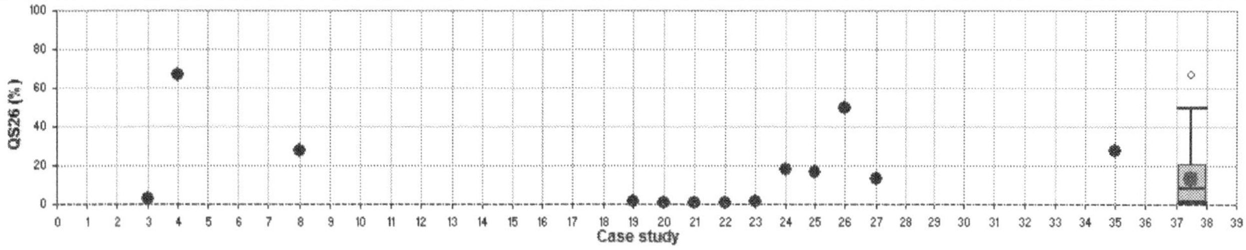

Comments: The mean value of the results collected is 16% and 50% of the case studies are situated in the range 1.2 to 18%.

Difficulties: Most end-users had difficulties with F15 (Customer complaints on interruptions) and F11 (Service customer complaints), informing that these data are not recorded. Two end-users recommend abandoning this indicator.

Relevance: This PI is considered important or useful by almost all end-users.

PI guidance range: The number of complaints depends much on cultural factors and therefore it does not seem advisable to establish a guidance range for this indicator.

QS26a - CRITICAL INTERRUPTIONS COMPLAINS (%)

Concept: Number of complaints due to supply interruptions to sensitive service connections during the year / number of service complaints during the year x 100

Processing rule: QS26a= F15a/F11 x 100

Variables: F15a – Customer complaints on critical interruptions (No.)

F11 – Service customer complaints (No.)

Parent PI: QS26

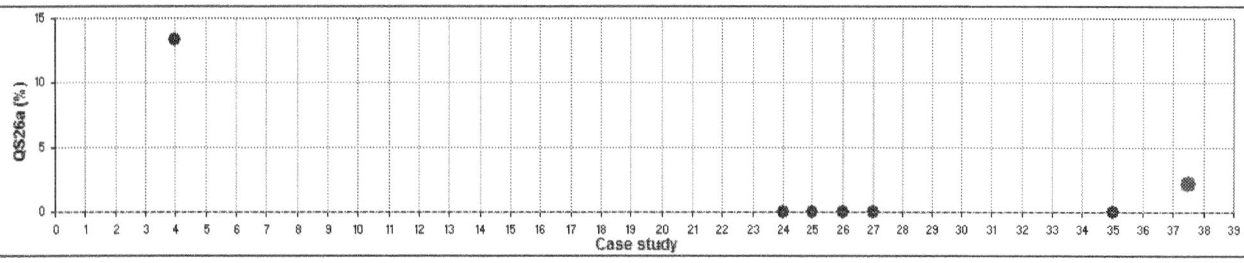

Comments: Only 4 results were provided, in a range of 0.00 to 13%. With such a small sample, no trends can be identified.

Difficulties: Most end-users had difficulties with the interpretation on critical interruptions and therefore with the assessment of this indicator, particularly with F15a (Customer complaints on critical interruptions) and F11 (Service customer complaints), informing that these data are not recorded. Two end-users recommend abandoning this indicator.

Relevance: This PI is considered important or useful important by most end-users.

PI guidance range: The number of complaints depends much on cultural factors and therefore it does not seem advisable to establish a guidance range for this indicator.

1.3 Financial Indicators

1.3.1 Annual costs

Fi1 - UNIT TOTAL COSTS (€/m³)

Concept: (Annual running costs + annual capital costs) / authorised consumption (including exported water)

Processing rule: Fi1 = (G2+G3)/A19

Variables: G2 – Annual running costs (€/year)

G3 – Annual capital costs (€/year)

A19 – Authorised consumption (m³/year)

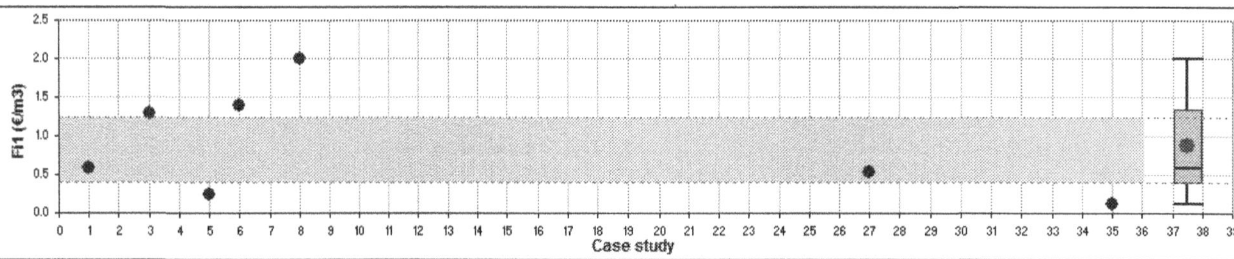

Comments: The mean value of the results collected is 0.88 €/m³ and 50% of the case studies are situated in the range 0.39 to 1.4 €/m³.

Difficulties: Several end-users had difficulties with G2 (Annual running costs) and G3 (Annual capital costs), informing that these data are not easily available. No end-user recommends abandoning this indicator.

Relevance: This PI is considered important by most end-users.

PI guidance range: At this stage, a tentative recommended guidance range seems to be from 0.40 to 1.25 €/m³.

Fi2 - UNIT RUNNING COSTS (€/m³)

Concept: Annual running costs / authorised consumption (including exported water)

Processing rule: Fi2 = G2/A19

Variables: G2 – Annual running costs (€/year)

A19 – Authorised consumption (m³/year)

Parent PI: Fi1

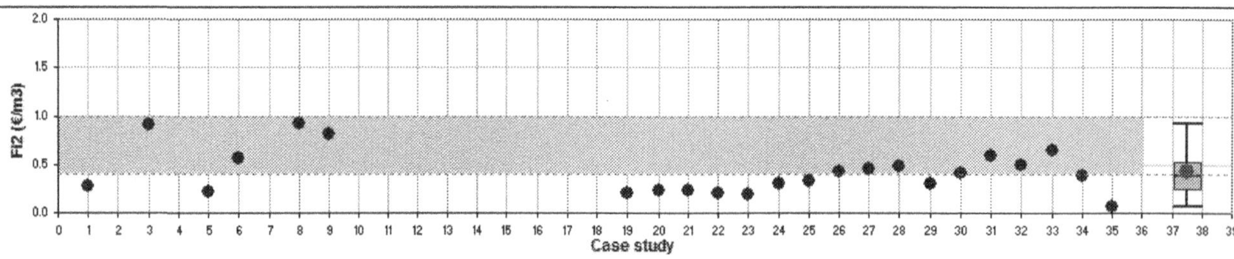

Comments: The mean value of the results collected is 0.43 and 50% of the case studies are situated in the range 0.23 to 0.55.

Difficulties: A few end-users had difficulties with G2 (Annual running costs), informing that these data are not easily available for specific networks of sectors. No end-user recommends abandoning this indicator.

Relevance: This PI is considered important by most end-users.

PI guidance range: At this stage, a tentative recommended guidance range seems to be from 0.40 to 1.00 €/m³.

1.3.2 Investment for network mains

Fi18a - UNIT INVESTMENT FOR NETWORK MAINS (€/m³)

Concept:	Annual cost of investments (expenditures for network mains) / authorised consumption (including exported water)
Processing rule:	Fi18a = G26/A19
Variables:	G26 – Annual investment in tangible assets (€/year) A19 – Authorised consumption (m^3/year)

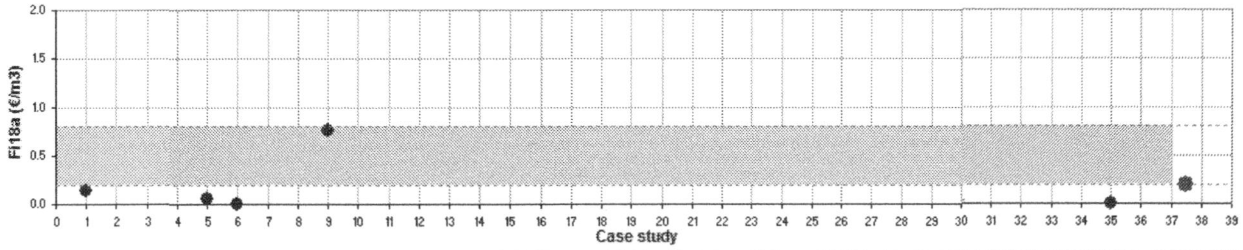

Comments:	The mean value of the results collected is 0.19 €/m^3 and 50% of the case studies are situated in the range 0.00 to 0.14 €/m^3.
Difficulties:	Several end-users had difficulties with G26 (Annual investment in tangible asset), informing that these data are classified or not easily available for specific networks of sectors. One end-user recommends abandoning this indicator.
Relevance:	This PI is considered important by most end-users.
PI guidance range:	At this stage, a tentative recommended guidance range seems to be from 0.20 to 0.80 €/m^3.

Fi19a - INVESTMENTS FOR NEW MAINS AND REINFORCEMENT OF EXISTING MAINS (%)

Concept:	Cost of investments for new mains and reinforcement of existing mains / total cost of the investments x 100, during the year
Processing rule:	Fi19a = G27/G26 x 100
Variables:	G27 – Investments for new mains and reinforcement of existing mains (€/year) G26 – Annual investment in tangible assets (€/year)
Parent PI:	Fi18a

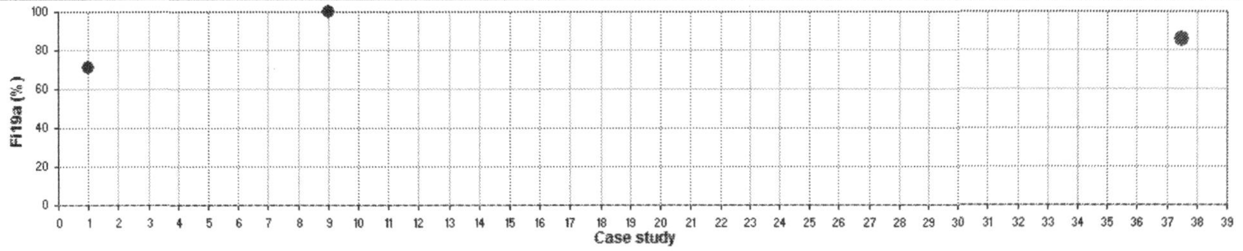

Comments:	Only 1 result was provided (71%). With such a small sample, no trends can be identified.
Difficulties:	Most end-users had difficulties with G27 (Investments for new mains and reinforcement of existing mains) and G26 (Annual investment in tangible assets), informing that these data are classified or not easily available for specific networks of sectors. One end-user recommends abandoning this indicator.
Relevance:	This PI is considered important by most end-users.
PI guidance range:	It does not seem to be advisable to define a guidance range for this indicator, because it depends much from case to case.

Fi20a - INVESTMENTS FOR ASSET REPLACEMENT AND RENOVATION (%)

Concept:	Cost of investments for the replacement and renovation ("like for like") of existing mains / total cost of the investments x 100, during the year
Processing rule:	Fi20a = G28/G26 x 100
Variables:	G28 – Investments for asset replacement and renovation (€/year)
	G26 – Annual investment in tangible assets (€/year)
Parent PI:	Fi18a

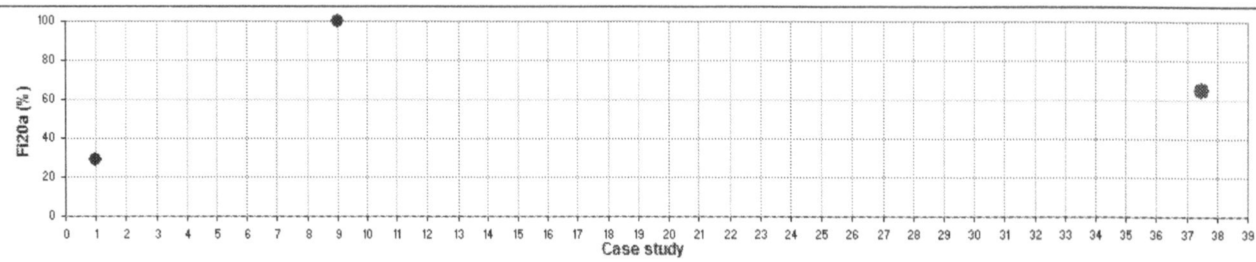

Comments:	Only 1 result was provided (29%). With such a small sample, no trends can be identified.
Difficulties:	Several end-users had difficulties with G28 (Investments for asset replacement and renovation), informing that these data are classified or not easily available for specific networks of sectors. One end-user recommends abandoning this indicator.
Relevance:	This PI is considered important by most end-users.
PI guidance range:	It does not seem to be advisable to define a guidance range for this indicator, because it depends much from case to case.

1.3.3 Tariffs

Fi21 - AVERAGE WATER CHARGES FOR DIRECT CONSUMPTION (€/m^3)

Concept:	Annual water sales revenue from residential, commercial, industrial, public, institutional and other customers (exported water excluded; public water taxes excluded) / (total annual authorised - exported water)
Processing rule:	Fi21 = G30/(A19-A9)
Variables:	G30 – Water sales revenue for direct consumption (€/year)
	A19 – Authorised consumption (m^3/year)
	A9 – Exported treated water (m^3/year)

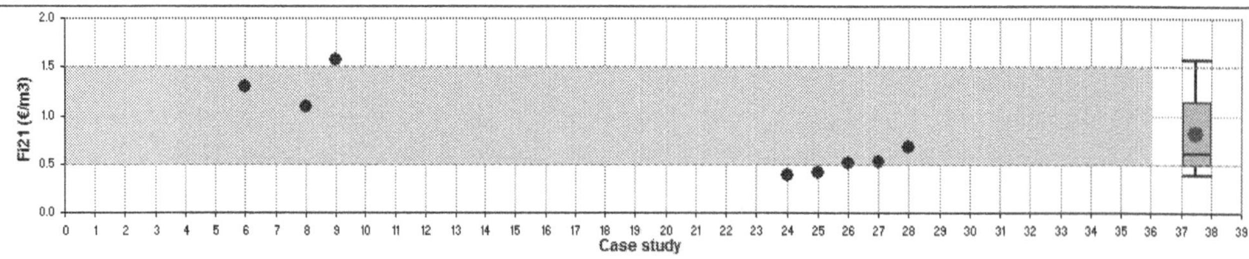

Comments:	The mean value of the results collected is 0.81 €/m^3 and 50% of the case studies are situated in the range 0.49 to 1.1.
Difficulties:	Several end-users had difficulties with G30 (Water sales revenue for direct consumption) and A9 (Exported treated water), informing that these data are available (without specifying the reasons). One end-user recommends abandoning this indicator.
Relevance:	This PI is considered essential or important by most end-users.
PI guidance range:	At this stage, a tentative recommended guidance range seems to be from 0.50 to 1.5 €/m^3.

Fi22 - AVERAGE WATER CHARGES FOR EXPORTED WATER (€/m^3)

Concept: Annual water sales revenue from exported water (excluding public water taxes) / exported water

Processing rule: Fi22 = G31/(A6+A9)

Variables: G31 – Water sales revenue from exported water (€/year)

A6 – Imported raw water (m^3/year)

A9 – Imported treated water (m^3/year)

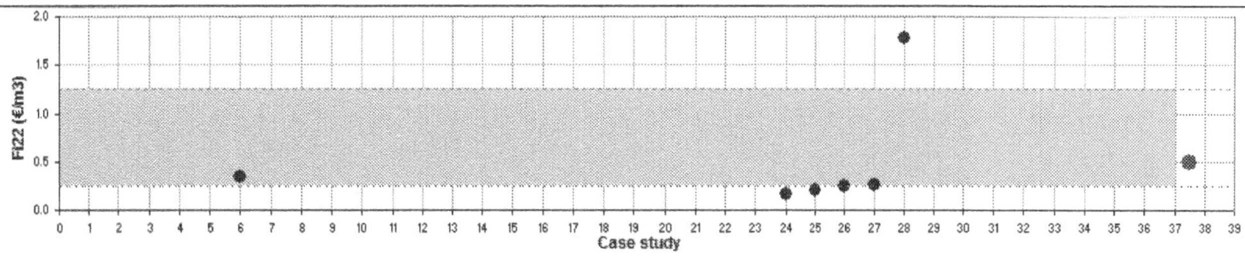

Comments: Only 6 results were provided, in a range of 0.16 to 1.7 €/m^3. With such a small sample, no trends can be identified. One case study presents a relatively high value of the indicator, caused by the preventive systematic replacement of old valves of unacceptable quality. All of those replacements were registered as failures.

Difficulties: Several end-users had difficulties with G31 (Water sales revenue from exported water) and A9 (Imported treated water), a few of them informing that these data are not applicable. No end-users recommend abandoning this indicator.

Relevance: This PI is considered important by most end-users.

PI guidance range: At this stage, a tentative recommended guidance range seems to be from 0.25 to 1.25 €/m^3.

1.4 Water resources indicators

WR1 - INEFFICIENCY OF USE OF WATER RESOURCES (%)

Concept: Real losses / system input volume x 100

Processing rule: WR1 = A24/(A4+A5+A8) x 100

Variables: A24 – Real losses (m^3/year)

A4 – Water abstracted (m^3/year)

A5 – Imported raw water (m^3/year)

A8 – Imported treated water (m^3/year)

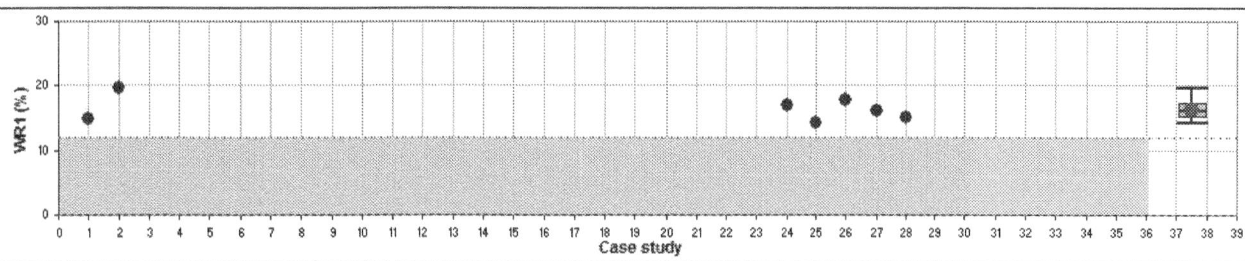

Comments: Only 7 results were provided, in a range of 14 to 28%. With such a small sample, no trends can be identified.

Difficulties: Several end-users had difficulties with A24 (Real losses), A4 (Water abstracted), A5 (Imported raw water) and A8 (Imported treated water), informing that these data are or not easily available. Three end-users recommend abandoning this indicator.

Relevance: This PI is considered useful or irrelevant by most end-users.

PI guidance range: At this stage, a tentative recommended guidance range seems to be less or equal to 12%.

WR2 - ANNUAL WATER RESOURCES AVAILABILITY RATIO (%)	
Concept:	System input volume / total annual abstraction capacity and imported water allowance x 100
Processing rule:	WR2 = (A19+A20)/(A1+A2) x 100 or WR2 = (A7+ A8)/(A1+A2) x 100
Variables:	A19 – Authorised consumption (m^3/year) A20 – Water losses (m^3/year) A1 – Annual abstraction capacity (m^3/year) A2 – Imported water allowance (m^3/year) A7 – Water produced (m^3/year) A8 – Imported treated water (m^3/year)
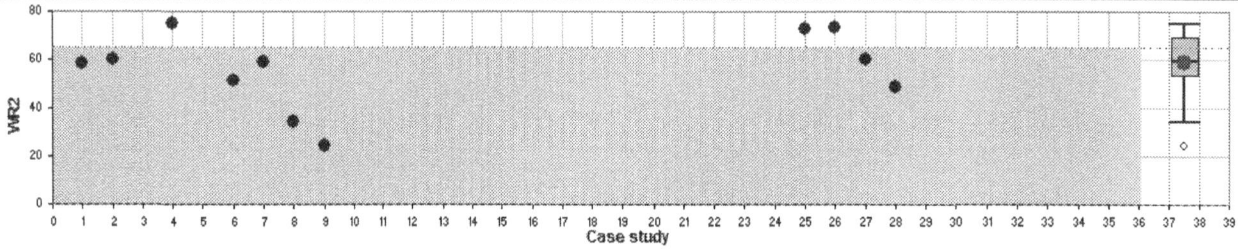	
Comments:	The mean value of the results collected is 56% and 50% of the case studies are situated in the range 50 to 67%.
Difficulties:	A few end-users had difficulties with A1 (Annual abstraction capacity), A2 (Imported water allowance), A7 (Water produced) and A8 (Imported treated water), informing that these data are not easily available. One end-user recommends abandoning this indicator.
Relevance:	This PI is considered useful by most end-users.
PI guidance range:	At this stage, a tentative recommended guidance range seems to be less than 65%.

1.5 Physical indicators

Ph3 - TRANSMISSION AND DISTRIBUTION STORAGE CAPACITY (days)

Concept:	Total capacity of transmission and distribution storage tanks (private storage tanks excluded) / [authorised consumption (including exported water) + water losses] x 365
Processing rule:	Ph3=C2/(A19+A20) x 365
Variables:	C2 – Transmission & distribution storage tanks capacity (m^3) A19 – Authorised consumption (m^3/year) A20 – Water losses (m^3/year)

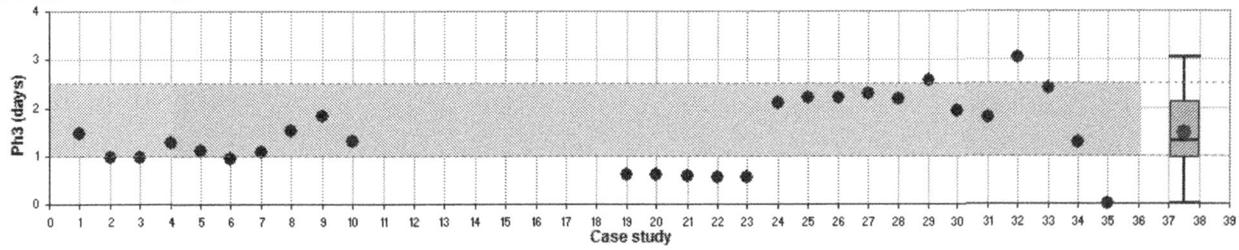

Comments:	The mean value of the results collected is 1.4 days of storage and 50% of the case studies are situated in the range 0.83 to 2.1.
Difficulties:	A few end-users had difficulties with A19 (Authorised consumption) and A20 (Water losses), informing that these data are not easily available. One end-user recommends abandoning this indicator.
Relevance:	This PI is considered essential, important or useful by all end-users.
PI guidance range:	At this stage, a tentative recommended guidance range seems to be from 1.0 to 2.5 days.

Ph7 - VALVE DENSITY (No./km)

Concept:	Number of isolating valves / total distribution mains length
Processing rule:	Ph7 = C30/C6
Variables:	C30 – Isolating valves (No.) C6 – Mains length (km)

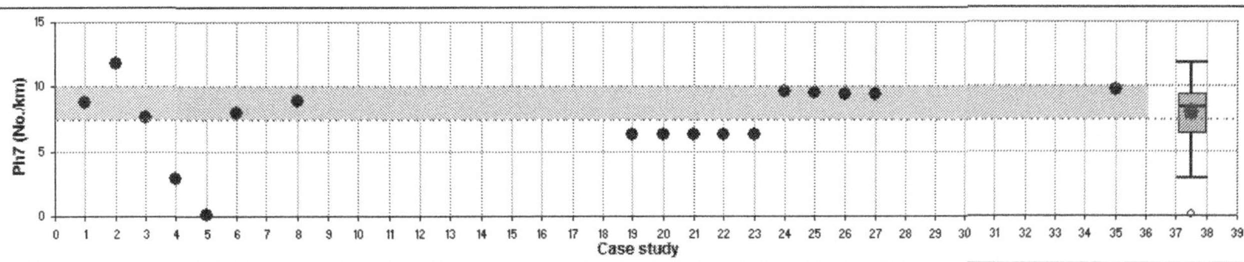

Comments:	The mean value of the results collected is 7.5 valves/km and 50% of the case studies are situated in the range 6.3 to 9.4 valves/km. Some case studies present very low results.
Difficulties:	Several end-users had difficulties with C30 (Isolating valves), informing that these data are not recorded. Two end-users recommend abandoning this indicator.
Relevance:	This PI is considered useful by most end-users.
PI threshold:	At this stage, a tentative recommended threshold seems to be from 7.5 to 10 valves /km.

IWA Publishing's authorised EU representative for General Product Safety Regulations is Diane D'Arras, 15 rue Duret, 75116 Paris, France, e-mail: safety@iwap.co.uk.

Printed and bound by CPI Group (UK) Ltd, Croydon, CR0 4YY
12/05/2026
02108893-0015